the joy of

keeping goats

SECOND EDITION

the joy of
keeping
goats
Second Edition

THE ULTIMATE GUIDE TO DAIRY AND MEAT GOATS

Laura Childs

the joy of

keeping goats

THE ULTIMATE GUIDE TO DAIRY AND MEAT GOATS

SECOND EDITION

Laura Childs

Skyhorse Publishing

Skyhorse Publishing books may be purchased in bulk at
special discounts for sales promotion, corporate gifts,
fund-raising, or educational purposes. Special editions
can also be created to specifications. For details, contact
the Special Sales Department, Skyhorse Publishing,
307 West 36th Street, 11th Floor, New York, NY 10018 or
info@skyhorsepublishing.com.

Skyhorse® and Skyhorse Publishing® are registered
trademarks of Skyhorse Publishing, Inc.®, a Delaware
corporation.

Visit our website at www.skyhorsepublishing.com

10 9 8 7 6 5 4 3 2 1

Library of Congress Cataloging-in-Publication Data is
available on file.

Cover design by Jane Sheppard
Cover photos: iStockphoto

First Edition ISBN: 978-1-61608-300-7
ISBN: 978-1-5107-1652-0
Ebook ISBN: 978-1-62636-942-9

Printed in China

To my first true love, Veronica. You are the phoenix.

CONTENTS

Preface   ix

Introduction   xi

①

Discovering the Joys of Keeping Goats   3

②

Getting Real about Getting Goats   17

③

Types, Breeds, and Strategy   25

④

Introduction to the Breeds   36

⑤

Getting Your Goats   54

⑥

Home Sweet Home   69

⑦

Pasture, Fields, and Forests   83

⑧

Feeding Goats   94

⑨

Goat Health   124

⑩

The Breeding Process   150

⑪

Goat Grooming Practices   179

⑫

Appreciating Goat Milk   191

⑬

Fiber Goats and Production Specifics   206

⑭

Favorite Recipes   215

Acknowledgments   223

Index of Tables   225

Goat Associations   227

References and Recommended Reading   229

Photo and Illustration Credits   231

Index   233

Preface

LIFE IN THE country. The place to raise children. To get back to the land. To learn that the soul of life is nothing more than lessons of decisions made and the personalities you've met along the way.

The young girl I handed my life over to some sixteen years ago will be leaving the farm. Off to college. Off to the city. I can barely breathe thinking of her returning to the very place I carried her from as a baby. I have secretly made plans to leave my haven, sustenance, and animals so that I may follow closely should she need me. It is a mother's prerogative.

I know myself too well though—in my heart I may follow, but my body will stay where the air is clean, and the animals I have come to rely on are right outside my door. Spoiled by country living for too long now, I can no longer hunt for my food in the concrete jungle.

Raising animals and living off the land prepares a person for these situations in life. There is a season for everything.

As one season passes, another will begin. I know it, but I am not quite ready to say good-bye to the summer storm that is my beloved Veronica.

We've learned so much here. We've learned of other people, of ourselves, and of what matters in life. For a while we thought we were pretty darn smart, until we learned that we know so very little indeed.

Introduction

ARE YOU DREAMING of becoming a goat farmer? When you picture that life in your mind, what does it look like?

I ask because there are varying visions regarding raising goats. Some of us dream of wandering lush green pastures bordered by snow-topped mountains with a herd of pristine dairy goats. Entrepreneurial spirits gaze over vacant land and dream in dollar signs after reading the research on the market demand for meat goats. Connoisseurs envision stainless steel pails of freshly gathered milk and all the potential cheese varieties they'll craft.

Those are just three of over ten ideals I hear regularly.

These diverse dreams of raising goats is not a human flaw. It is the fault of the goat! If goats weren't so incredibly well suited to perform so many tasks, we humans could get together and help each other discover the peaceful, rich, or creatively delicious life we yearn for.

In the meantime, we have joy. Joy in keeping them, in learning about their care, and in discovering their individual personalities.

And for your adventure into keeping goats, I wish you practicality as every farm animal you tend will require some knowledge, some common sense, and some compassion to successfully keep.

The content within the pages of this book will give you a good head start on turning your goat farming dream into a reality. Your goats will teach you the rest.

the joy of

keeping goats

① Discovering the Joys of Keeping Goats

WHAT'S NOT TO love about goats? They will feed you, clean up the overgrown mess in the fields, and take long hikes with you while carrying your supplies. They will make you laugh when you're sad, provide extra income for even the smallest farm, carry you to town and back in a little cart, and perhaps best of all, they will gaze upon your face with earnest adoration.

Most of us though, are excited about the prospect of keeping goats for their deliciously sweet milk and low-fat, nutritious meat.

Goats' milk is consumed in larger quantities by more people, than cows' milk. A staggering 65 percent or more of the world's population choose goat over cow's milk for reasons other than availability or economics. Perhaps even more surprising, over 60 percent of all red meat consumed worldwide is goat meat. Goat meat has substantially less than half the fat of chicken, beef, pork, or lamb.

To keep, the goat is both a pleasure and all-round multitasker. Traits that are well appreciated as we seek the means to cut back on stress and find joint purpose for every facet of our lives.

Goats provide many blessings for minimal cost and care and can easily be kept on just a few acres of land. They are a livestock staple in all of the old-world European countries and are finally gaining the attention they deserve from North American farmers.

The goat has been called the poor man's cow for far too many years. Although their service may initially fit the bill for the frugally minded, there is little chance of counting them as a lesser animal once you start tabulating their virtues. You may have one main reason for your interest in goats today, but once they are in your keep, you can't help but notice how wonderfully suited they are to a farm or family's needs. In time you will have forgotten why you bought that first goat, being left only to wonder how you ever got along without them.

Your appreciation of the goat won't stop at the barn door though. Every goat you pour your time into will also be working their goat-given charms on you. I have not met one goat keeper, even those with

Goats have been a barnyard staple in European countries for centuries, but we are only just now realizing their multiple uses and gifts.

large herds who hasn't been completely enamored with or emotionally connected to a favorite *Capra aegagrus hircus* (Latin: goat).

As you reap the gifts they repeatedly bestow, you'll wonder why you waited so long to discover the joys of keeping goats.

The Most Important Reason to Keep Goats

I am often asked which feature of the goat is the most valuable. The answer isn't mine to give. It is different for every person, on every farm, large or small. You may have a lactose intolerant child or become concerned about the safety of consuming commercially raised meats. Perhaps you've eaten some delicious chèvre (goat cheese) and want to try your hand at creating some of your own?

BELOW: Lovable goats make great pets and projects for children of all ages. The bond between the two forms quickly, simply with time spent playing in the yard or field.

As I look upon my own fields of this centuries-old farm, once covered in brambles and weeds from years of neglect, I might say the most important reason to keep goats is brush control.

Keeping the weeds down and the fields open for grazing doesn't just increase the value of a farm, it also affects the quality of life on that farm. Our region is predisposed to large populations of black flies and mosquitoes. Our goats keep the fields open to sun and wind, which in turn pushes those nasty spring and summer pests back to the forest interior.

If I were looking over my financial records (instead of my fields) when you asked that question, I'd say that the most important reason is economical. Raising goats has decreased the stress on our household budget while helping our family to be more self-sustained. We've cut back on our trips to town to purchase milk and cheese. We haven't needed to buy fertilizer for the garden for the past ten years. And our freezer is full of delicious meat at one half the cost from previous years.

Goats also make great pets and provide the lessons of responsibility for children. Your child might choose to care for and show a goat as part of a 4H project. If you homeschool like I have with Veronica, goats also provide good fodder for lesson plans—animal anatomy, how to make soap, or train a goat to walk on a leash, to name just a few exercises and studies we've performed in the past.

Throughout the years I have talked to hundreds of goat owners across North America who have responded to "Why do you keep goats?" with diverse replies. Here are the most common:

- Goat milk and meat is a superior option to grocer's beef and dairy.
- My goats keep the weeds under control in our fields while grazing with other livestock.
- The cost of raising goats pales in comparison to end-user purchase of milk, cheese, and meat.
- They are just a lot of fun!
- Our children wanted some after visiting a neighbor's farm. We've had goats on our land ever since.

- Home made cheese! Goats make the best milk for cheese, and we can enjoy it without unnatural preservatives, chemicals, or harsh processes.
- Goats are "green" livestock. They're environmentally friendly. Less methane gas per pound than a cow.
- They are our pets with a purpose. Our working companions on the farm.

Whatever your reason is for considering goats, I hope you find answers and encouragement within the pages of this book. I can say with near certainty that once you've raised a few, you'll never want to be without them again.

Who Are You?

For every noble human trait there will be at least one good reason to keep goats.

- If you are health-conscious, the goat provides lower fat meat and a higher-nutrient dairy product.
- If you are frugal, you'll enjoy the cost savings that goats afford the household budget.
- If you have a discerning taste, the goat offers delicious, made-for-your-palate cheese.
- If you have a maternal instinct, annual offspring genuinely reciprocate your doting attention.
- If you have a mind for business, current market demand will appeal to you.
- And should you just need a reason to laugh every day, the antics of a few goats will provide a lot of comic relief.

Our First Foray into Keeping Goats

"The goat becomes the professor whenever teachers can't be found," Turkish saying.

My daughter Veronica and I began raising goats eleven years ago, entirely by accident. This method of foraying into goat husbandry is not suggested, but it does illustrate how just about anyone can raise a goat.

BELOW: Goats are a forgiving and easy keep. Although our very first seemed beyond much hope, when we bought her she went on to grace our barn for many years and gave us many kids to love.

We were greenhorns, city slicker imports with seventeen empty acres and a barn to fill. This is why, while at the county auction to buy turkeys, the five-year-old Veronica saw a large white goat walking calmly on a leash and instantly fell in love.

Knowing absolutely nothing of goats, I could plainly see that her new object of affection was aged, tired, and had suffered neglect. Her hooves were so overgrown she walked on the folds of them. Her head hung low, and she would not make eye contact. Knowing this would be more of a rescue mission than a proactive step into goat ownership, I entertained the idea nonetheless. If you have read any of the last ten years of writing, you already know that Veronica is, and always has been, rather determined.

Ten minutes later and twenty-eight dollars poorer, we were the new owners of a tired goat.

I don't ever want you to buy a goat by following in these footsteps. There were clearly three strikes against this purchase—first, it was on the whim of a child; second, the goat was plainly ill-kept; and third, coming from an auction house, you never know what disease a goat could be bringing onto your land.

As you'll read in the chapter on goat health, a goat from a questionable home could bring many hardships to your farm. Ills that can outlast the life of that one goat.

We were fortunate. Nanny (Veronica named her) brought only blessings. Once she was in top form for her age and cleared by our veterinarian, we bought her a few goat companions. In return for her health and happiness, she gave us six years of milk, seven kids, a lot of laughter, and a fine excuse for an education in goat ownership.

Goats Reign Supreme

I am biased, I know it, but I also truly believe that goats are the animal of choice for farmers in the know.

Dare I say that goats are superior? They eat the weeds and brush that no other farm animal will eat, they supply a premium grade of milk with multiple uses, and their meat is of the highest caliber. Even the goat's excrement has a superior quality—it is 2.5 times richer in nitrogen and phosphoric acid than cow manure, making it an excellent choice for both vegetable and flower gardens.

When compared to cow milk, goat milk is higher in phosphorous, riboflavin, niacin, calcium, and vitamins A and B1 and is lower in cholesterol. Right off the udder, goat milk is loaded with antibodies and has a much lower bacterial count than cow milk. Based on these benefits, goat milk is often recommended by doctors to safely treat physical conditions such as vomiting and insomnia in infants, children, and pregnant women.

As if that wasn't enough proof of superiority, a goat will produce more milk than a cow from the same quantity of food. The nutrient conversion efficiency for milk production from a goat is 45 percent, the cow is only 38 percent efficient.

Goat milk is also richer than cow's milk. As you'll read in the following chapter, some dairy goat breeds produce milk with an average 6 percent butterfat, which may even jump above 10 percent during different stages of their lives. This makes goat milk an excellent choice for cheese production. The high butterfat content also lends itself to making luxurious soaps. Soaps that are often prescribed for treatment of skin rashes, minor irritations, and eczema.

Although some goats produce milk high in fat, the meat from that goat is leaner than chicken. Found most often for sale within ethnic and health-conscious communities, chevon or cabrito (meat from young goats) is growing in popularity. At the time of writing, North American–grown chevon is not meeting market demand. Most of our goat meat is imported at high cost to the consumer and to the environment.

Goats will give you much more than milk and meat. Aside from breeding your goats and selling live kids, you can also sell or use the hair from your goats. Goats produce two fibers—angora and cashmere.

With adequate fencing and fields to roam in, goats become an easy and inexpensive keep. Young dairy kids not only beautify the landscape, they will also provide milk for your family for years to come.

Angora is only grown by one breed, but cashmere is produced by many. Cashmere is the lightest weight, warmest, and completely nonirritating fiber known to man. One cashmere-producing goat will create a quarter to a third of a pound of cashmere every year.

If you're both business and environmentally minded, you'll enjoy this ancient business practice made popular again. The United States is seeing a revival in the business of renting goats out to land owners to control overgrowth. Construction and telecommunications companies have been securing the assistance of small herds to clear vines and brambles from building sites and from the bases of cellular and radio transmission towers. Forty goats can clear 3,600 square feet of land per day, and they do it all without expensive machinery or greenhouse gas emissions.

Goats: So Much More Than Grins and Giggles

Goats can be both sweetness and shenanigans, loving and demanding, obedient and troublesome—all within 130-inch frame. They can double you over with laughter as easily as they make you cry out in utter frustration.

To keep them safe while maintaining your own sanity, remember my favorite mantra: "A goat is like a three-year-old in a goat suit."

Goats' antics, quick minds, and ability to find trouble where mere mortals had never thought to look make caring for them a lot like running a day care for preschool children—really smart preschool children. Don't let the challenge put you off though. Much like raising children, with a little experience, you'll quickly learn to stay one step ahead of them.

Our first doe taught me a lot about the intelligence of a goat. For her own reasons that I would never understand, she would only leave the barnyard area when the front door of the house was open. This was usually on our arrival home from town. The moment we'd come barreling down the driveway with a car full of groceries, she'd begin her barnyard escape. By the time we were returning for the second carry, there she was, standing in the middle of the kitchen.

ABOVE: Sweetness and shenanigans. One good goat can keep you laughing every day with their crazy antics and silly behavior.

Over the years, her occasional appearance inside the house was endearing. The morning that she awoke a world-renown speaker and business client—who had driven for ten hours for a private consulta-tion and then spent the night in the front bedroom—was not. While I was in the hen house collecting eggs for breakfast, Nanny took the opportunity to sneak in and nuzzle the cheek of my sleeping city-slicker guest.

(2)

Getting Real about Getting Goats

Basic Goat Anatomy

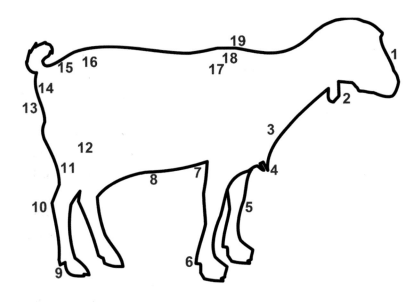

Key

1. BRIDGE OF NOSE

2. DEWLAP

3. POINT OF SHOULDER

4. BRISKET

5. KNEE

6. DEW CLAW

7. HEART GIRTH (TO WITHERS)

8. BARREL (TO BACK)

9. PASTERN

10. HOCK

11. FLANK

12. STIFLE

13. ESCUTCHEON

14. PINBONE

15. THURL

16. HIP

17. CROP

18. SHOULDER BLADE

19. WITHERS

Body Condition Scoring Assessment

Once you have a few goats in your barn, you will be able to tell a lot about their mood and well-being at just a glance. Healthy goats have shiny coats and are curious by nature. They hold their head, ears, and tails up and have a firm, confident stance.

We all know that we can't judge a book by its cover though. To truly assess a goat's condition, whether you're considering a purchase or adjusting feed quantities, you'll want to perform a body condition scoring assessment.

Body condition scoring is a standardizing system developed for all animal types. Goats are no exception. Use the chart below to perform a hands-on assessment for goats.

BODY CONDITION SCORING ASSESSMENT

Score		Spinous Process	Rib Cage	Loin Eye
BCS1	Very thin	Easy to see and feel the sharp point of bones can be felt.	Well pronounced, easy-to-feel top of, and slightly under, each rib.	No fat covering.
BCS2	Thin	Easy to feel but smooth.	Smooth and slightly rounded. Slight pressure is required to feel individual ribs.	Smooth with a slight, even fat cover.
BCS3	Good Condition	Smooth and rounded.	Smooth and even to a gentle touch.	Smooth, even fat cover.
BCS4	Fat	Can feel with firm pressure, but no points can be felt.	Individual ribs can not be felt, but you can still feel a slight indent between them.	Thick fat cover.
BCS5	Obese	Smooth; no individual vertebra can be felt.	Smooth to the touch. No individual ribs can be felt. No separation of ribs felt.	Thick fat covering. The fat may feel lumpy.

The spinous process is in reference to the vertebra of the spine. When you first touch the goat for assessment, run your fingers down the spine from the shoulders to the tail head and assess the feel of the bones and padding between hide and bone.

Next, touch and assess the rib cage on both left and right sides, from the top of the back to the under side of the goat.

Finally, stand behind the goat and position both of your thumbs on the spine just behind the start of the rib area. Curl your fingers down and back as though you were going to pick the goat up using just your hands on the muscle there. You now have the loin area cupped in your hands. This will be the most difficult area to assess until you have practiced on goats of differing condition.

More advanced assessments can be performed, but these are difficult to learn without assessing many different goats. For the best visual representation of body condition scoring, visit the Langston University research website at http://www.luresext.edu/goats/research/bcs.html

Goat Speak—A Few Terms You Need to Know Now

Throughout this book and in your journey of discovering more about goats, you'll run across new terminology. Here are the most common.

- Dam—A goat's mother.
- Doe—A female goat. Young, unbred does are called doelings.
- Buck—A male goat. Young bucks are called bucklings.
- Brood Doe—A doe kept for breeding purposes, usually possessing excellent genetic traits or ancestry.
- Chevon—Meat from mature goats.

- Cabrito—Meat from goat kids.
- Cud—The regurgitated food of a ruminant.
- Herd—Two or more goats.
- Kid—A newborn goat. They are called kids until they reach a year of age.
- Ruminant—Any animal with four stomachs that chews and regurgitates their food.
- Sire—A goat's father.
- Teat—The two protrusions from the udder that milk flows through.
- Udder—The organ of a doe that produces milk.
- Wether—A castrated male goat.
- Yearling—A goat between one and two years of age.

BELOW: Side by side, it isn't that hard to tell which one is the goat and which one is the sheep, but sometimes it isn't all that easy.

Telling Goats From Sheep

Do you have trouble telling the difference between goats and sheep? It is more common than you think. Since I was once a greenhorn desperately trying to get up to speed in country culture, I learned the easiest way to tell the difference between the two. It's all in the lips; goats have a solid upper lip, sheep have a divided one. Here are some other ways to tell the difference.

GOAT OR SHEEP?

Goat	Sheep
Sometimes bearded	Never bearded
Eats by browsing and grazing	Never browses, always grazes
Dislikes getting wet	Does not mind rain or snow
Curious personality	Shy personality
Tail stands up	Tail hangs down
Independent in action and thoughts	Always moves with or follows the flock

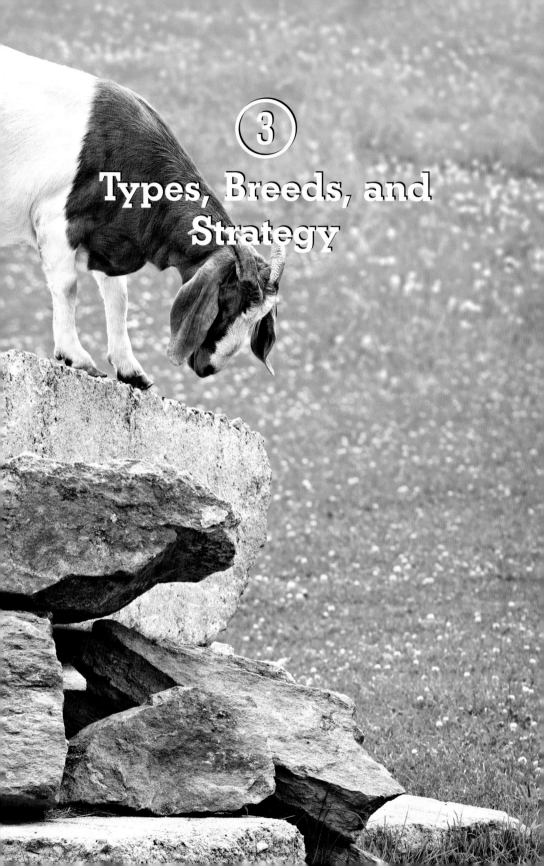

③ Types, Breeds, and Strategy

IF YOU ARE fastidious in preliminary research, you can choose a breed this month and happily raise that breed for many generations. That noble path is the road least taken. Of my contacts, an estimated seven out of ten new goat owners changed their preferred breed within the first year. For this reason alone, starting with just a few goats for the first year is prudent.

There are over two hundred registered goat breeds worldwide. Eighty of these are registered for agricultural use. Only twenty or so are commonly available in North America. Fewer still may be suitable for your region's climate.

Pure goat breeds are separated into three main groups for farm use—milk, meat, and fiber. Dairy and fiber breeds can be raised for the secondary purpose of meat. As for secondary purpose, a few of the meat goat breeds produce marketable fiber, and both fiber and meat goats could be milked. Not one breed, however, is versatile enough to be considered excellent at all three. Choose a breed based on the trait most important to you and consider any secondary ability a blessing.

Two main factors can make or break your success and enjoyment with keeping goats. The first is that the goats you keep suit your needs. The second is that they should suit your personality. This may seem strange to a new hobby farmer, but the personality factor is a crucial consideration.

As a gross generalization, some breeds exist happily in the field with only once daily interaction and minimal chore time. Others yearn for, and will demand, much more attention. Goats, left alone to their own devices, will return to the feral state quickly—second only to domestic cats left in the wild.

Once you begin visiting goats for purchase, you will see varying personality types and mannerisms within the breed. Although the range is not as apparent as canine types, there is one important parallel. As long as you are not working against a basic instinct (i.e., survival, feeding), you can mold a type or trait with training to suit your style.

Although nearly every well cared for goat fits my criteria for selection, you may have other ideals. I look for inquisitive and attentive

eyes, a medium to high level of physical energy, and a willingness to please. When evaluating kids, you will need to make your assessment based on the doe's disposition or clues the breeder gives you regarding the kid's start in life.

The standard breeds you'll read about below have similarities in more than just looks. Their mannerisms and types are also similar. The mixed breed or experimental goat—which account for over 60 percent of all goats in North America—will almost always possess the best traits from both side of their heritage.

Experimentals are the created offspring of a managed cross-breeding strategy. These goats are often hardier than their purebred dam or sire—they just can't always be registered within their dam or sire's goat registry. As well as others, the Pygora and Kinder breeds where once classified as experimentals, but today they enjoy a registry of their own and full-breed status across the continent.

You will also come across goats that are labeled as grade or scrub. These goats are much cheaper to purchase and are a perfectly acceptable meat goat. A grade doe might even supply ample milk for a family's needs. They are no less valuable than their purebred counterparts once you've proven their capabilities.

Average Yield for Full-Size Registered Breeds

- Dairy—The average dairy doe supplies 900 quarts of milk per year; 2–4 quarts per day.
- Meat—The average buck kid provides 25 to 40 pounds of meat. Stockier meat breeds such as the Boer provide twice as much.
- Fiber—Adult Angoras supply 10 to 15 pounds of mohair per year. Adult cashmere producing goats might supply one-third of a pound per year.
- All—Supply approximately 1 pound of garden enhancing manure per day and excel at weed and brush removal.

Your Goat-Keeping Strategy

Before we discuss the various breeds, let's first explore a few farm strategies. Defining your strategy is as important, if not more important, than choosing the breed to suit your needs.

First of all, you should have at least two goats. Goats are herd animals; and they truly are happiest when they have one of their own kind to mingle with.

In most strategies you'll want two does, two wethers, or one of each. Keeping a buck or a breeding pair never fares well for goats. Full (unneutered) bucks produce a strong-tasting meat that is uncomplimentary for the North American palate. If a buck is housed with a doe, their presence will taint the delicate taste of milk and scent the hair of the doe. More importantly, if you keep a buck with a doe in pasture, you will lose control over all aspects of breeding.

At the very least I ask that you have at least one barn buddy to keep your goat company. Goats make great companion animals for

a horse or cow. They also pair up nicely with sheep, but since their mineral requirements differ, sheep should never be allowed access to the goat's mineral block.

At any rate, if one goat is a charming addition to your farm, two will be twice the fun. So let's explore the options of keeping at least two goats and how it might benefit you. We need to first consider the output of each goat and how you'll put that output to use for your benefit. These are small farm or backyard farming strategies; if you plan on having a large herd, you likely have other ideas on management and production.

The Oberhasli breed are known for their gentle natures and beautiful markings.

Goats Earn Their Keep

- Dairy—Aside from the obvious option of selling your dairy goats' offspring once or twice a year, you could make and sell goat's milk, soap or lotion made with the milk, or delicious artisan cheeses for local restaurants or markets.

- Meat—The market for goat meat is growing across the continent. In 2007, Canada imported over 3.5 million pounds of goat meat. The United States imported over 22 million pounds. Estimating an average carcass weight of 32 pounds a piece, we can approximate that 800,000 goat carcasses were brought onto North American soil for the year.

- Fiber—Goats produce mohair and cashmere used in high-end clothing and accessories. The majority of market production of mohair in the United States originate from Texan farms. Creating a sizable income from the fiber market alone requires a substantially sized herd and specialized knowledge. The secondary meat market of less productive animals is strong.

- Pets—Goats can be great pets for both adults and children. The standard-sized breeds are common in the pet market, but the popularity of pet miniature goats is increasing. These pint-sized personalities can easily be kept with little more than a large backyard to explore. Should you decide to breed and sell miniature kids, you'll find prices comparable to those of a purebred puppy.

RIGHT: Don't overlook the flashy Nigerian Dwarf goat as being just a pretty face. They are excellent milkers, have the most multiple births, and produce a high-quality milk that is perfect for making delicious cheese.

ABOVE: The Boer breed grows fast and has excellent mothering ability. Boers are currently the most popular of the meat breeds in the United States and Canada.

Main Goal: Dairy

Raise dairy goats with the added benefit of meat once per year.

By purchasing two bred dairy does, you will have your first taste of farm-raised milk as soon as the kids are born. If you stagger their breeding each year, say four months apart from each other, you'll keep the milk flowing for your family.

Doelings could be kept to increase herd size or sold as potential milkers for other farms. Bucklings could be sold to others to raise for meat, butchered before weaning, or grown on for another six months at minimal cost.

Main Goal: Meat

One of the most popular strategies involves breeding dairy does annually to keep them in milk. The kids are then raised for just a few months when they can either be sold or butchered and consumed.

The second most popular is to purchase weaned kids in early summer to raise for meat. By the time winter rolls around, your freezer is full of low-cost but highly nutritious meat. Purchase two young wethers for this strategy. If available in your area, get Boer or Boer-cross wethers for the highest yield.

When you're raising kids for meat, your goal is to raise the largest possible kid, in the shortest period of time, at the lowest possible cost. The least expensive goat meat to raise is from a six- to eight-week-old kid. Since dairy kids will only weigh about thirty-five pounds at that age, you will only net fifteen to twenty pounds of freezer meat. If time, money, and space permits, you could grow him on for a larger yield. By twelve weeks of age, he will weigh fifty pounds and will have only cost you a few dollars in grain and hay. Put him on pasture (with normal supplemental feeding) and by the time he's reached seven to eight months, he should weigh in at eighty pounds; about thirty-five pounds for the freezer. These are all estimations based on a 45 percent dressing percentage of goats.

The exceptions to the rule of age-weight increase are the Boer and Angora breeds. Boers are larger at birth and grow faster than any other goat in the same amount of time. The twelve-week-old kid example from above could easily be twice the weight of a dairy or grade kid at the same age. Angoras grow at slower rates than the meat or dairy breeds and do not reach their full weight until their second birthday.

Main Goal: Fiber

If you'd like to raise goats for fiber and have no interest in raising kids for meat or milking does daily, two or more wethers are ideal. Wethers are very friendly as a rule and yield a higher fiber count than does.

You could also purchase two bred Angora or cashmere-producing does to have a little milk, the fiber you desire, and offspring that eventually feeds your family.

④

Introduction to the Breeds

Of Dairy Breed Interest . . .

- Nubians produce milk with high butterfat content and are often referenced as the Jersey equivalent of cattle. They are also considered to be the most vocal.
- Saanens and Sables produce the highest volume of milk. They are compared to Holstein cattle. Their milk is lower in butterfat than the other breeds, which makes it more palatable to unconverted fans of cow milk.
- The Toggenburg goat is the oldest registered breed of any animal in the world, with records tracing back to the seventeenth century.
- If you are more interested in milk than the meat from the offspring, consider a miniature version of your favorite dairy goat; almost all breeds are available as minis today!

Dairy Goat Breeds

Of the six common dairy goats, the Swiss breeds (Alpine, Oberhasli, Saanen, Toggenburg) are the hardiest for colder climates. The remaining two (LaMancha and Nubian) are genetically equipped to handle extremely warm and dry climates but may be kept in the north with proper care.

The majority of dairy goats in this country are managed much the same as dairy cows. They require milking twice daily and have a 305-day lactation cycle. This allows a 60-day dry period used by the doe to centralize her energy into the growing fetus and to replenish her body's store of nutrients. When a doe has her kid (freshens), it is common practice to remove the kids and raise them separately from their mothers.

Alpine

The Alpine is one of the larger dairy goats and a popular breed for commercial dairies. Discovered in Switzerland, the Alpine breed

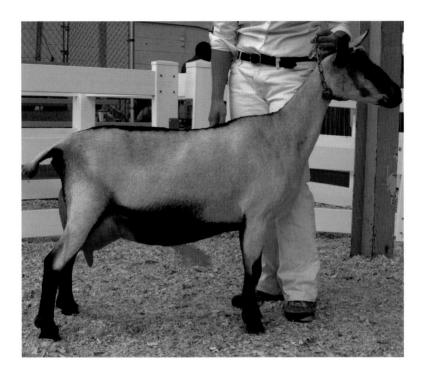

gained quick favor across Europe. Alpines match the Saanens in milk quantity.

This breed is known for her amiable personality. She has a straight or slightly dished nose and erect ears. Hair is medium to short. Most often seen in variations of black and white or brown and white. Other color patterns include chamoisée (any shade or mixture of brown, often with a black stripe along the back and white markings on the face), two-tone chamoisée (usually a lighter color on the forequarters), cou Clair (a light-colored neck), cou blanc (a white neck with black rear quarters), cou noir (black front quarters and white hindquarters), sundgau (black with white facial stripes, white below the knees and hocks and white on either side of the tail), and pied (broken with white, spotted, or mottled).

Alpine does should be at least thirty inches tall at the withers and weigh at least 135 pounds. Bucks should be at least thirty-two inches tall and weigh at least 160 pounds.

Also referred to as the French, Swiss, British, or Rock Alpine.

LaMancha

The most striking feature of the LaMancha breed is ear formation. Ear shape is rated as either a gopher ear or an elf ear and should be no longer than two inches by breed standards. Gopher ears contain no cartilage, just a ring of skin around the auditory canal. Elf ears contain a small amount of cartilage and a small amount of skin, which could turn either up or down from the cartilage. For buck registration, only gopher ears are allowed.

The origin of the LaMancha is a controversial subject amongst breeders. Some say this breed is a direct descendant of a breed imported to the United States from Mexico in settlers' times; others say they originated in Spain. No concrete documentation of their origin has been found to date. The LaMancha breed was officially recognized in early 1950s in the United States.

LaManchas have consistently pleasant temperaments and are considered by many to be the calmest of the purebreds. Hair is short, fine, and glossy and can be any pattern, color, or combination of colors. They are top producers of high butterfat content milk.

LaMancha does should be at least twenty-eight inches tall at the withers and should weigh at least 130 pounds. Bucks should be at least thirty inches tall at the withers and should weigh at least 155 pounds.

Nubian

The Nubian goat is the most popular dairy goat for commercial dairies. Considered to be the most vocal and energetic of the breeds. Nubians will thrive in almost every North American climate with some consideration for extreme temperatures and weather conditions. Their milk is high in butterfat (5 percent or more) and are an excellent choice for small farms interested in cheese making.

The Nubian breed originated in the United Kingdom and is popular with both British and Near-East cultures.

The most striking feature of the Nubian are their large pendulous ears, which are both long and wide. They have a strong convex profile. Hair is short, fine, and glossy. These moderately large goats are found in many shades and marking patterns of brown.

Nubian does should be at least thirty inches tall at the withers and weigh at least 135 pounds. Nubian bucks should be at least thirty-two inches tall and weigh at least 160 pounds.

Also known as Anglo-Nubians

Oberhasli

Oberhasli was once classified and registered with the Swiss Alpine and American Alpine breeds. Although not as common as some of the other dairy goats, this breed is growing in popularity across the United States due to their stunning appearance and a calm, gentle temperaments.

Their heads are broad at the muzzle, tapering to a thin nose. Foreheads are wide with prominent eyes. Ears are short and erect and point forward. Face shape can be dished or straight. This breed is most commonly found in hues of reddish brown with accents of black along their dorsal, legs, belly, and face. On occasion, a fully black kid will be born from a breeding of two Oberhasli.

Oberhasli does should be at least twenty-eight inches tall at the withers and weigh at least 120 pounds. Bucks should be at least thirty inches tall and weigh at least 150 pounds.

Also known as Swiss Alpine, Oberhasli Brienzer.

Saanen and Sable

Saanens and Sables are registered separately but differ only in the color of their coats. The Saanen is white, the Sable taupe. This large dairy breed produces an average of four to six quarts of low butterfat milk per day.

Saanen and Sables have forward-facing, slightly erect ears. Faces and noses are narrow. Profiles can be dished or straight. Both sexes are bearded. Amiable, gentle, and calm natures.

Saanen and Sable does should be at least thirty inches tall at the withers and weigh at least 135 pounds. Bucks of these breeds should be at least thirty-two inches tall at the withers and weigh at least 160 pounds.

Toggenburg

The Toggenburg is the oldest registered breed. The Toggenburg breed is small in stature and produces a low butterfat milk (2–3percent). With pleasing personalities and a small frame, this goat is relatively easy to manage.

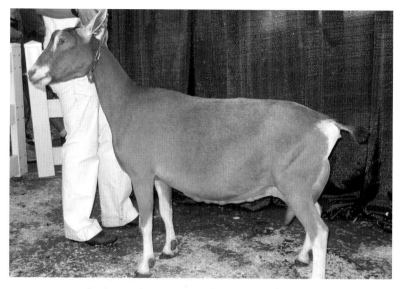

Coloration ranges from light fawn to a dark chocolate. Toggenburgs have white markings, white ears with a dark spot in middle, white stripes from above each eye and extending to the muzzle, white socks from hock to hoof, and a white triangle on both sides of the tail. Wattles and beards are common for both sexes. Profile can be straight or dished.

Toggenburg does should be at least twenty-six inches tall at the withers and weigh at least 120 pounds. Bucks should be at least twenty-eight inches tall and weigh at least 145 pounds.

Meat Goat Breeds

Quick Facts on Goat Meat

- Australia exports over sixteen thousand metric tons of goat meat annually. Over 50 percent of that export comes directly to North America.

- Goat meat is the most common red meat consumed worldwide. North American market demand is growing. Increased ethnic diversity represents the greatest demand for goat meat but health conscious consumers are quickly discovering the many uses for this lean and tender protein source.

- Cabrito, the meat of a ten- to twelve- pound milk-fed kid, is a considered a delicacy among the Hispanic culture. The United States Hispanic community is currently over thirty-five million strong. This community is forecasted to comprise 18 percent of our total population by 2025.

- African, Asian, Caribbean, Filipino, Jewish, Italian, and Middle Eastern cultures also purchase goat meat for standard dinner fare as well as religious and ethnic holidays.

- The Roman Easter market prefers goats that weigh 20 to 50 pounds. The Greek market, between 55 to 65 pounds. The Muslim market, 50 to 70 pounds and 100 to 115 pounds. The Christmas market has the widest range: 25 to 100 pounds.

Although any breed of goat (including the grade or scrub goat) can be used for meat, this classification is specific to the few breeds who grow the quickest and have more lean muscle volume than other classes.

The Spanish and Myotonic goats have been raised for centuries by North American farmers to provide a reliable meat source for their families. Both grow to decent proportions and are well muscled animals.

In less than twenty years (beginning in 1993), three new meat goat breeds have entered and dominated the North American market. The South African Boer breed, the New Zealand Kiko goat, and the Savannah breed. Although the Boer has seen the most favored, the Kiko is gaining ground due to their excellence in high meat-to-bone ratio. Savannah goats are relatively new to the scene, but interest among southern goat farmers is strong.

Boer

Boers were developed in South Africa to grow quickly, produce flavorful meat, clear overgrown land, and thrive in inclement weather. This thick-boned and heavily muscled breed is capable of

ABOVE: Purebred Boer Buck, CSB Ruger Reloaded Ennobled

grazing during all seasons and will grow larger in less time than any other breed.

Boers are commonly seen with white bodies and reddish brown heads, but solid varieties and spotted types are rising in popularity. The Boer is mild tempered and easily trained.

Boer does range from 200 to 250 pounds, bucks from 240 to 320 pounds.

Kiko

The Kiko is the product of a New Zealand government-funded initiative that began in the 1970s. The crossbreeding program involved both wild and domestic goats with the intention of producing a fast-growing meat source. Within a few generations and only minor variations during development, the Kiko was established. Kiko translates as "meat for consumption" in Maori (the native tongue of New Zealand).

Kikos are now proven to have the highest occurrence of kidding twins, a fast and reliable growth rate, and an unmatched aversion to disease.

ABOVE: Purebred Kiko Buck, Penn Acres Huckleberry

A Kiko may not grow as heavy for the scale as the Boer but it does have a higher meat to bone ratio, therefore providing more consumable product per pound. Kikos are generally white but color variations can also enjoy full-blooded pedigree.

Kiko does average 100–150 pounds, bucks from 250–300 pounds.

Spanish

The Spanish goat has been present on American soil since the days of early exploration. These goats were either left behind or escaped and became feral. Through evolution they have grown climate tolerant and well muscled. For these reasons, the Spanish goat is often used in crossbreeding programs across the country.

Between current crossbreeding practices and numerous feral generations, the Spanish goat is found in all color variations, face shapes, and weights. The most notable characteristics are horn curvature and ear carriage. Ears are long and wide like a Nubian ear, but carried horizontally and slightly forward. The breed is currently a conservation priority by the American Livestock Breeds Conservancy.

ABOVE: Purebred Savanna Buck

Spanish goats are often incorrectly labeled as scrub, brush, or grade goats (terms used to define goats of no known heritage).

Spanish does grow to an average of 150 pounds, bucks to 250 pounds.

Savanna

The Savanna goat is the newest introduction from South Africa. This breed is hardy and sun-tolerant with a temperament similar to the Boer. Popularity and interest is growing for this new-to-the-scene breed.

Savanna goats are usually pure white with solid black hides. With a roman nose (rounded face) and long pendulous ears, they resemble a Nubian in the head shape, but a Boer in body type.

Savanna does grow to an average of 130 pounds, bucks to 240 pounds.

Myotonic Goat

The Myotonic goat has a genetic propensity to myotonia congenita. Myotonia is a muscular response to a startle. The response results in a spasm that locks the animal's legs, potentially causing unbalance similar to a human fainting. Brain and nervous system activity is unaffected by the myotonic response. This response condition also renders the animal helpless in the pasture until the animal regains control.

Myotonics were introduced into America by a Canadian (Nova Scotia) breeder. Repeated stiffening has given this breed muscular thighs—enough to be classified as a meat goat. Raised more as a novelty in modern times, the Myotonic breed was listed in the threatened breed category by the American Livestock Breeds Conservancy until just recently.

This is an intelligent, friendly, and easy-to-keep breed of goats with minimal desire to climb over or jump fences. Available in a wide variety of colors and coat lengths, these goats are straight-faced with long, slightly protruding ears and bulging eyes.

Myotonics are the smallest breed in the meat class (seventeen to twenty-five inches tall at the withers). Does should reach 130–150 pounds. Bucks have been noted to reach 200 pounds.

Also known as Wooden Leg, Stiff Leg, Nervous, or Tennessee Scare Goat.

Fiber Goats

Two distinct fibers are produced by goats—mohair and cashmere. While mohair is only produced by the Angora goat, cashmere can be found on over sixty of the world's goat breeds. (In North America, we most often find cashmere on Spanish and Myotonic breeds.)

Finding a top-producing cashmere goat is difficult in North America, not to mention an expensive acquisition for the small herder. Primarily you won't be looking for a breed, but a type (or class) of goat. For a goat to be classed as a cashmere, it should produce fiber with a crimp, be under 19 μ in diameter, and over 1 1/4 inches in length. The size of μ, from a simplistic understanding, is the result of a calculation from an equation of two or more measurements.

Angoras, on the other hand, are a pure and registered breed. These silky-haired goats generate ten to fifteen pounds of mohair annually. A wether produces slightly more than a doe.

The Angora is certainly in a class of its own, but the care is similar to that of other goat breeds. If you are only planning on keeping a few and hope to sell the fiber twice a year, find a local breeder or Angora Goat group to collaborate with. Not only will they share invaluable breeding, feeding, and coat care tips, they will also assist you in finding a buyer for your yield.

Angora

An import from Turkey, the Angora breed has been in North America for over 150 years. Texas is the second largest producer or mohair worldwide. The breed is neither hardy nor an easy keep in comparison to other breeds, but they are an amiable pet for knowledgeable and conscientious owners.

Angoras are a silky-haired breed with long coats of wavy or curly locks, five to six inches in length. Although most Angora goats kept for fiber collection are white, they are also found in black, red, and brown.

Angoras are not a particularly hardy breed. As focus is on coat growth and quality; their diet should be closely monitored. Prone to parasitic infection given their dense coats and seldom set out to browse in mixed pasture where seeds and burrs can collect in the coat, lessening the value.

Does grow to an average of 95 pounds, bucks to 190 pounds.

Miniatures

Miniatures are currently popular on farms with limited space or need for full-production animals and their output. When I decided a few years ago to settle on a breed, I almost passed over any goat labeled as a mini, dwarf, or pygmy until a goat-keeping friend convinced me to consider the yield and personalities of these breeds.

Although only two miniature breeds are widely available in the United States and Canada, more than a few crossbreeds are worthy of a second look.

The Pygora (a cross between an Angora and a Pygmy) produces a lesser grade mohair but will also produce the fine down a Pygmy provides and finish larger if you are interested in meat. The Kinder (a cross between a Pygmy and a Nubian) is a great dual-purpose cross-bred for both milk and meat.

Finally, almost every dairy breed has a miniature version now. If you run a search on the Internet you'll find Mini Manchas, Mini Toggs, Mini Nubians, and more. Many of these miniature versions provide a better match of production quantities for the average family!

The miniature breeds are perfect for family farms, make great pets, and are easy enough for children to handle. These goats are 1/3 to 2/3 the size of a full-size breed and require much less space and feed.

Nigerian Dwarf

Considered both a miniature and a dairy goat, the Nigerian Dwarf doe can provide up to a gallon of milk per day at her peak, but she averages a little over a quart by standard. Small- to medium-sized families interested in putting high-quality milk and cheese on the

ABOVE: Purebred Nigerian Dwarf, Hidden Hollow Wysteria

table would be hard pressed to find a better match than the Nigerian Dwarf. Butterfat content in milk ranges from 5 percent on average to 10 percent near the end of lactation.

The Nigerian can also breed year-round and is known for having multiples—twins, triplets, and quadruplets.

The face of the Nigerian Dwarf can be straight or slightly dished. Their coats are straight and medium length and come in every pattern and color imaginable. Most are naturally horned and some will have blue eyes. They are friendly, easy to work with, and bond quickly with their keepers.

By breed standards, Nigerian Dwarf does should be at least seventeen inches tall at the withers, bucks at least nineteen inches and neither should be taller than twenty-one inches. Mature does average thirty to fifty pounds, bucks and wethers forty to eighty pounds.

African Pygmy

The Pygmy Goat, another African import, has gained wide popularity as backyard pets due to petting zoo exposure. They are hardy and adaptable and make excellent little browsers for overgrown fields and small pastures.

Although it isn't documented, there are two breeding styles for Pygmy owners. While some will breed for the purpose of pet stock, others are bred to be milked. You will need to have small hands to milk a Pygmy, but at her peak, a doe can produce two quarts per day of a high butterfat milk (approximately 6 percent). The milk is higher in calcium, iron, potassium, and phosphorus than the average dairy breeds' milk and lower in sodium.

The small stocky legs and lack of agility of the pet Pygmy makes containment less challenging than other breeds. Coats are straight and of medium length and come in a variety of patterns and colors.

By breed standards, does should be no taller than 22 3/8 inches at the withers, bucks no taller than 23 5/8 inches. Weight averages are fifty to eighty-five pounds.

BELOW: Purebred African Pygmy, Monterey Bay Equestrian Center

⑤ Getting Your Goats

I Just Want One

PLEASE GET TWO. You could have more than two, but you should have at least two. Every goat needs a companion.

Having witnessed multiple testimonies regarding single-goat ownership I agree that some goats are *capable* of living alone. The question, however, is not if they are capable, but if they will thrive in your care.

Nature dictates, by the very essence of the beast, that a goat should be part of a herd. Your goat has a long ancestry of wandering plains, mountains, and deserts with at least a few of her own kind. A horse, cow, or ewe will make a fair companion animal, but to pair them up this way is to stray from thousands of years of instinctual programming. You are not ever likely to see see a goat partnered with a horse or cow, wandering together in the wild.

Animals that are designed by nature to live in a herd will experience stress when forced to live without their own kind. Internalized stress often manifests as illness, compromises immune systems, and lowers production. Stress will also affect longevity. Goats need goats.

Where to Find Your Goats

If a particular goat hasn't already caught your eye and captured your heart, it's time find the best stock that suits the goat shed you've created, the role you'd like them to fulfill, and your budget.

One of the best places to find local breeders with third party opinions is your feed supply store. The staff or owner of your local store know the people who purchase goat ration and medications. They should be genuinely happy to help you out; you are a potential new customer after all. Be sure to elicit their opinion on the breeder.

You want to find the person in your area who has the best goats and is a knowledgeable and attentive owner.

Goat owners are fanatics. We love to talk about goats, our favorite breed, other people's goats, and to help out new goat owners. Good owners are happy to give advice on all aspects of care at any hour of the day or night. Each one that you meet can open up an entirely new network of nearby breeders and goat husbandry associations.

While you are at the feed store getting a referral, check the bulletin board for advertisements. If you don't see a goats-for-sale sign, be proactive and write up a quick "Goats Wanted" ad and post it to the board before you leave. On your way out the door, look for a farm newspaper or magazine distributed by, or available at, your feed store.

Your goats could be just one click away! Over the last few years, the Internet has grown enough to easily find nearby breeders or associations for every breed. Type your state or province, plus the breed you seek into a Google browser window. You should find hundreds of listings.

Goats need other goats. If one is a charming addition to your barn, just imagine how much fun two can be!

Five years ago I bought two prized horses as the result of an online search. I also made a new friend as a result. We e-mailed, passed photos back and forth, and haggled on the price, all without actually seeing the horses in person. After one visit to the farm, I knew I was dealing with a hands-on breeder who kept excellent care of her animals, and the deal was sealed. If I can buy horses (of a breed that are at risk of endangerment) online, you will surely find your goats there.

Other viable options are your local newspaper classifieds, livestock auctions, and county fairs.

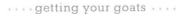

At the fairs, you'll find livestock shows where the owners are present. Ask about upcoming stock they might have for sale and exchange phone numbers or get a business card if they have one.

The auctions are another great place to network and learn if you're the friendly type. On occasion, you may luck out and find a beautiful animal at a decent price. I once picked up a gorgeous doe with a buckling at her side at less than market rate. All the goat buyers kept their paddles down once they saw me place a bid. Later in the crowd, I heard a boy about ten years of age whisper to his mother that the money didn't matter, his goats were going to a home and not a butcher. It seems they brought the pair to auction simply because they didn't have time to wean, disbud, and castrate that little buckling.

Until you are experienced at picking out healthy stock and know all the potential dangers of bringing a sick or infested animal home, leave your pickup truck and livestock trailer at home. If you bring it to auction with you, you'll be tempted to fill it up with animals that other owners are trying to move out. These animals, often just a step away from being culled at home, are not the best option for a new goat owner.

Most sellers that you meet are happy to arrange a private showing at their farm at a later date. Trade phone numbers and give them a call in a few days. Networking is a golden opportunity to make new friends who share your interest, and you might end up purchasing a special goat that a seller was apprehensive to sell to "just anyone."

Viewing a goat on their own farm gives you the opportunity to observe housing arrangements, discuss registration, ancestry, and temperament, and view barn records for the goats you're interested in.

The Best Goat for Your Money

There's more to purchasing a goat than price and paperwork. Health, age, production (current and in ancestry), and temperament are all key considerations. Whether you're buying purebred registered stock or crossbred goats, a responsible owner can quickly answer most of your questions regarding birth date, veterinarian visits, and ancestry. He will have documentation on last worming dates, blood tests, vaccines, and any medication used throughout his ownership of the animal. The barn will be clean or at the very least, not fragrant with ammonia.

The goat you consider should be easily handled and exhibit neither shyness nor aggressive tendencies. They will be as interested in you as you are in them. If a doe isn't looking at you with shining eyes and an air of curiosity, move on to the next potential.

Know your breed. The body type should be a fair match to the breed you have chosen. Some goats have wide faces, some are without ears, others have height ranges they should have met by maturity. In all goats, look for a wide and strong back and chest, straight legs with trimmed hooves, and a clean shiny coat. Avoid any goat with a sway back, a pot belly, bad feet, or a defective mouth.

The only way you'll develop an eye for healthy productive goats is to closely observe as many of them you can. Whether you subscribe to a goat magazine, spend time at county livestock judging, discuss conformation with local breeders, or surf the breeder sites online, you will soon build up enough of a breeder's eye to distinguish the difference between a strong healthy animal and an unproductive underperformer.

Ask to see the dam and, if possible, the sire too. By all means get your hands on any goat you are considering. Perform the body condition scoring assessment that you learned in the previous chapter

ABOVE: Viewing goats on their own property will give you a better idea of their personalities, their past care, and their ancestry.

before investing your heart and finances. Look for signs of external parasitic infestation. (See Lice in the chapter on Goat Health.)

Allow personality to play a role in your decision. You will be spending some time with your goats, so make sure you get along with each other. You can tell a lot about a goat's personality when you handle them. If you can't easily touch a goat for assessment before the point of sale, you won't be able to manage that goat on your own property.

Steer clear of the extremes. Goats purchased for brush control or to grow for meat will still need to be caught and handled for routine care. If you come across an erratically active or painfully shy goat, you might want to take a pass until you have more experience with the nature of a goat and are up for a challenge. Gaining the trust of a seemingly wild or shy animal can be a great blessing, but it requires great skill and patience.

Goat Speak: More New Terms You Need to Know

On the day you buy your goat, you might hear some new terms.

- Registered Purebred—Comes with a traceable pedigree (much like a purebred dog comes with registration papers and a family tree).
- Advanced Registry—Pertaining to milk does. This goat has been noted and registered as supplying a decent volume of milk over the course of a year. Dependent on her current age and health, Advanced Registry does have a proven record of milk production.
- Star Milker—Pertaining to milk does. The star system is based on a one-day test by volume (see sidebar).
- Grades—A grade goat may or may not be a purebred animal. It is without papers and registration. If your goat meets certain requirements and you desire it, you might be able to register it as a "recorded grade" with the issuing authority in your area.
- American, Experimental, Native on Appearance (NOA)— These are also grade goats. Americans and Experimentals are the result of crossbreeding programs and are not purebred animals, but a NOA might be.
- Polled—A goat without horns. The goat may have been born that way or had the horns removed later in life.

The Grade Debate: Do You Need Papers?

Although registration offers some reassurance that the animal you're buying is of notable heritage, it is not to be mistaken as a guarantee of production. A registered doe with an impressive ancestry may

not be such an impressive milker. She could have problems producing enough milk to keep the barn cat interested, have trouble kidding, or both.

Registration papers matter most to those who plan to show, breed, or otherwise profit from the animals. If your strategy of keeping goats is for personal use or meat, the added expense and paperwork (both now and in the future) might just be a waste of your resources.

Although my view isn't popular, I contend that registration of a farm animal is similar to the registration of a purebred dog—an unnecessary expense for the average person. You can look at a pure-bred dog and know that it is purebred without seeing the paperwork and general assumptions of production, personality, and growth rate based can then be made. For personal or farm use a registered dog or goat is paperwork shoved in a desk drawer. unless you are planning on breeding, showing or selling registered animals.

Somewhere on your adventure of keeping farm animals you'll discover your own comfort level or need for paperwork and registra-tion. At the end of the day, no amount of registration and paperwork beats trusting in both the seller and your own ability to assess the age, health, and potential of an animal.

Bringing Your New Goat Home

Any change in a goat's surroundings and routine will cause stress. Know your goat's current feed program (right down to the very hour) and bring a week's supply of her previous ration and hay home with you. For a few days, don't alter her old routine, then slowly switch her over to your farm's hay, grain, and any supplements. This is true whenever you are making changes to a goat's feed or routine, make the change gradually over the course of a few weeks.

The seller should supply you with the following:

- registration papers (if applicable)
- veterinarian contact information
- list of past medications and vaccinations
- feed (grain, ration, and hay) for the first transitional week
- hooves trimmed and horn buds removed (if applicable)

BELOW: Although this doeling looks shy peeking out behind the tree, she is just playing coy. As long as she'll let you catch and hold her at this age, you should have a good match for your farm.

Most sellers will worm the goat twenty-four hours before you pick her up. This ensures that she does not introduce more worms to your land. If the seller has not wormed her, do so a few days into quarantine on your own farm.

Kid goats can be transported in a pet carrier or dog kennel in the backseat of your car. I have seen people transport full-size goats in the backseat of the family car, but I wouldn't do so unless the trip was twenty minutes or less and your route accommodated slow driving.

The back of a pickup truck with a cap is perfect for transporting goats if you don't own a livestock trailer. Add three to four inches of straw, cover any metal loops or clasps in the box, and you're set for the ride home. Take it easy on the turns, curves, and bends in the road.

You'll probably want to stop and check on your new goat after every turn in the road. Twenty minutes into the ride, and I'm a bundle of nerves. After half an hour, I'm stopping to check. The goats we've transported have always been fine when I open the truck hatch, staring back at me as if to say, "What's the holdup? Let's get going."

Kids can be wrapped in a towel and transported on your lap if they are small enough.

If you live more than ninety minutes away, I suggest borrowing a trailer or hiring a livestock hauler to transport the goat.

Once home, plan on quarantining all new goats for thirty days. During the first week she will be adjusting to new sounds and surroundings. The worming medication will have worked its way through her system, and she will start settling in.

During chore time, the quarantined goat should be attended to last. You'll be entering her stall with your other goats' scents, and you won't be potentially transferring any germs to your existing herd.

Pick up some disposable painter's slippers from your hardware store and slip on a new pair every time you enter the quarantine area. I started doing this when I was visiting chicken farms to ensure I didn't bring diseases home with me, then adopted it as a habit when bringing new animals in as well. Disease is often spread by feces and bodily fluids.

Call the vet in for a checkup, booster vaccinations or blood work. If you can schedule your new acquisition's arrival with other veterinarian work needed, you'll save the cost of a farm visit.

Ensure all test results are back and cleared from the veterinarian before you release the goat on your property or into an existing herd. By the time the quarantine is over, you will have spent ample one-on-one time together and your goat will be bonded to both you and your property.

Star Milker System Explained

The star milker system is a series of calculations on the three factors of milk production: quantity, duration, and quality. If a doe earns eighteen points at the end of the calculation, she'll receive one star. If her dam or grand-dam also had a star, she can earn two more and be labeled as a three-star milker. A lofty and valuable title indeed!

Point Calculation:

Factor 1: Total pounds of milk produced in one twenty-four-hour period (1 point per pound).

Factor 2: Total number of days in lactation (.1 percent for every ten days). To a maximum of three points.

Factor 3: Butterfat content (percent x points/pounds, divided by .05)

As an example:

Factor 1: Your doe gives 6.0 pounds in morning and 7.5 pounds in the evening. She earns 13.5 points.

Factor 2: She's been lactating for the last 42 days. 42 x .01 earns her .42 points.

Factor 3: Her butterfat percentage was 3.8. She would gain 10.26 points based on this equation: 3.8 x 13.5 = .513 divided by .05 = 10.26

Factor 1 + Factor 2 + Factor 3 = 24.18

In the example above she will have earned one star.

Once a star has been earned a doe can inherit more from her ancestry. (e.g., a doe has earned her own star and her dam and grand-dam had been registered star milkers, then that doe would be considered a three-star milker.)

6

Home Sweet Home

IN THE WILD, goats adapt and survive in all climates, often in less than ideal conditions. Your goat, even after centuries of domestication, is almost (but not quite) as hardy. A little shelter and a lot less stress goes a long way for an animal, doubling their lifespan and production rates when compared to their feral cousins.

To house and confine a goat safely, you must first understand their abilities, their mind-set, and their desire. All goats are inquisitive by nature. The largest order of the day is to protect them from sampling or getting tangled up in the things that can cause them harm.

Remember the adage of a goat acting like a preschooler in a goat suit. If a barrier can be jumped over, an electrical wire touched, glass windows to push upon, grain to be tasted, or nails to be stepped on, it will certainly be challenged, broken, eaten, chewed, ripped, pushed, and punctured by a goat. If you wouldn't leave your three-year-old nephew alone for twenty minutes in the shelter or hope to hold him with the fence you just built, it probably isn't adequate for a goat.

BELOW: "A fence that won't hold water, won't hold a goat." This new fence looks sturdy but wherever a goat can push a head through eventually her body will follow. Within no time at all, she's knocking on your front door, asking you to come out and play.

Goats won't take up much room on your farm. Aside from their feed bins and a small platform to sleep on, their housing requirements will be minimal. In many cases, a large shed will work fine for a few goats. With ingenuity and a little room in your budget, you can have the ideal set up for keeping goats.

Housing Systems

Oh Those Poor Freezing Goats!

Dairy goats will be at their best when temperatures are between 55 and 70 degrees Fahrenheit. Milk production and feed consumption levels experience noticeable changes only in the extreme temperature ranges—above 80 degrees and below 0 degrees Fahrenheit. With only a few concessions in care the goat is capable of coping in temperatures below freezing.

Dairy goats don't sweat. Dependent then on the climate in which you raise them, the challenge is often greater to keep them cool in the summer than it is to keep them warm in the winter.

There are two primary methods to housing and containing goats. The first is to pasture them and provide a poor weather and bedding shelter. This is method of choice for the hardy meat breeds or for large herds of dairy goats from spring until autumn. In regions where the weather dips well below freezing for days on end, goats will need a little more protection.

If the pasture area is too far away from their full service shed or barn, you'll need to provide a pasture shelter. This doesn't need to be much more than a 4' to 6' high, three-sided enclosure with a slanted roof. Allow 8' to 10' square feet per goat in the pasture shelter.

The other method, loafing and confinement, is the act of keeping goats in a shed or small barn with a fenced yard for exercise.

The loafing and confinement system of raising goats is used mostly for dairy breeds or by farmers who don't have ample pasture. Sufficient room is provided for the goats, both inside and out, for sleeping, eating, fresh air, sunshine, and exercise.

The act of confinement keeps high energy activity to a minimum. Less energy expended by the animal allows for more productive use of feed. Angoras nearing a sheering date are often kept this way to control parasitic infestations and organic matter getting caught in their long hair.

The average dairy goat requires only twenty square feet of living space indoors, plus two hundred square feet of exercise space outdoors. Meat goats require more: thirty square feet inside, three hundred outside. Large herd meat producers only provide fifteen square feet inside as the animals prefer to be on pasture for all but the worst weather. Miniatures require a third less than dairy breeds.

I personally wouldn't confine a goat to such a small amount of space, but I have the luxury of acreage. I have had friends who didn't have much space but gave their goats a very nice life nonetheless, and I'm sure you could do the same if the situation called for it. Fresh air, cleanliness, clean water, and compassion make far happier lives than twice the space and an uncaring owner.

Shelter From The Storm

Goats are happy to spend their days and nights on pasture with the bare minimum of cover. Most breeds are comfortable between temperatures of 40 and 70 degrees Fahrenheit, but none of the breeds like to be wet. Meat breeds may continue to graze in a light summer rain but should have a shelter to return to at all times. Forced exposure to rain in low temperatures often results in illness. Kids and low-condition individuals are at the highest risk for respiratory infection and hypothermia.

Extreme Weather Temperatures

Not one of the domestic breeds common to North America are built to handle long bouts of below-freezing temperatures. These goats, unlike other livestock, do not grow enough of a thick coat or put on a layer of fat directly beneath their hide to prepare for the

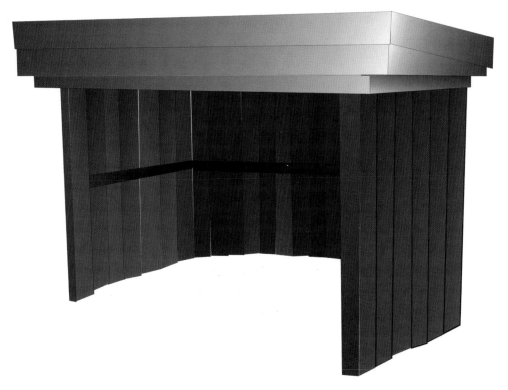

ABOVE: An outdoor sheltered feeder keeps supplemental hay off the ground and protected from the elements.

full thrust of winter. They will keep themselves warm by huddling together and through energy expenditure through rumen activity. Adequate shelter, herd mates, and sufficient hay make the coldest winter nights bearable.

Dairy goats that are used to being locked in for the night should be kept in draft and moisture-free buildings with ample dry bedding. Take extra care during extremely cold temperatures if kidding is imminent or if you already have kids in the barn due to early-season births. Extra bedding, a supervised or safe heating unit, and a little extra hay to eat will help keep the harsh chill from bothering your herd.

Meat goats that are kept on pasture year round should have their shelter turned in such a way that the prevailing winds aren't blowing into the open side of the shelter. More bedding should be added (daily if necessary) to ensure that the goat's extremities aren't touching frozen ground. More hay will also be required to replace the energy spent on staying warm.

If at no other time, cold days and nights are the perfect time to treat your goats to warm drinking water. Goats actually prefer slightly warm water in all seasons so the investment of safely installing an immersion heater or a water bucket with the heater core built right in would make the cold season more liveable.

During dry or hot summer months you'll find goats equally resilient and capable of managing in extreme temperatures. They may move around less, spend more time in the cool shade of the barn, and drink more water to stay comfortable. If you lock up your goats at night, you'll want to make an exception for the hottest nights unless you have plenty of windows that provide a cross breeze.

With all the extra time spent in barns or shelters, litter and bedding need to be changed more often. Goat owners can't afford to miss the extra work required during these times. Urine and feces quickly build high ammonia levels, which can cause respiratory health complications. Kids are at the highest risk as they are closer to ground level where ammonia is strongest.

Converted Sheds

If you will just be raising a few goats and you already have a shed on your property, you might be able to convert it to a nice little barn for your goats. Conversions save you time and the cost of building a brand new structure. Knowing the number goats you will house at the height of the season (your does plus offspring that you keep for a few months) will determine if an existing building is adequate.

When I decided to move my dairy does out of the communal barn and into their own space, we took an old 10' x 12' shed and spruced it up instead of starting from scratch. Fresh wood went on the walls, and a small slab of concrete was poured onto the milking area floor. After a few partial walls, a manger, and a sleeping platform were added and the goats moved in.

In an 10' x 12' shed, the communal stall will take 35–50 percent of your floor space. This leaves adequate room for a milking station, feed storage and one or more smaller stalls. You'll need at least one small stall for isolation of a sick goat, quarantining a new goat, kidding, or weaning.

Shed Conversion Example

Two dairy does given 40 square feet (20 square feet each) for a communal stall, plus a smaller stall for kidding (20 square feet), a section for milking and grain storage (35 square feet), plus 20 square feet extra for two kids annually.

At less than 120 square feet required, a little planning can convert a 10′ x 12′ shed into an adequate loafing barn. Tight, but adequate.

If you're building from scratch, design your floor layout to accommodate the feeding and watering of goats without entering their communal stall. The easiest way to do so is to build a half wall between their space and yours. Your side contains the manger, water bucket, and soda/salt feeder. Their side contains slatted or key-holed head access to all three. Open areas (above the half walls for example) can be securely sealed with 12″ x 12″ page wire fencing material.

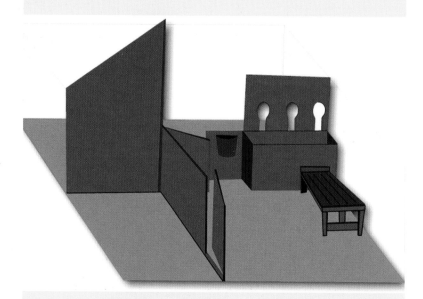

ABOVE: A converted 10x12′ shed. Showing a milking stand, keyhole manger access, exterior water buckets, and good distribution of space: 60 square foot communal stall, 20 square foot kidding stall, and 35 square feet of work and storage area. Add an access door to the back wall that opens to their yard and you're all set to bring your goats to their new home!

Indoor Mangers

Designing and building a slatted or key-holed access manger may be extra work right now, but it will save you money in the long term. Goats are not only notoriously picky about the hay they eat, they are also the most wasteful. If the manger is open, you will soon discover that goats take a mouthful of food, turn to see who might be behind them every few minutes, and drop half of their mouthful on the floor. Once it has hit the floor, they will have very little interest in it.

Goats also love to climb into the top of open mangers. They will stand there and eat, curl up and fall asleep, trample the hay with dirty hooves and perform body elimination functions. Once soiled, no self-respecting goat will touch the hay.

ABOVE: Two different styles of head-only access for feeders. The one on the left is traditionally made of plywood; on the right, rough 2 x 4 lumber.

The standard top width of a keyhole is eight to nine inches with a keyhole shaped taper to the bottom at four to five inches wide. The height of the keyhole from top to bottom is sixteen inches. When the opening height is optimal, goats will crane their necks to put their head in at the top and then lower their heads to a comfortable fit within the slot.

If the top of your wall is higher than most of your goats can reach through slats or keyholes, you could build a variable height platform on their side of the wall.

Keyhole entries won't work for horned goats. They won't be able to fit their heads in, and if they somehow manage to do so, they could get stuck and panic when they are finished eating.

A slat-access manger built high on the wall will work for both polled or horned goats. Width between the slats should be about six inches—just enough for a goat's nose to push through and grab a mouthful of hay. If you can build it with a hinged rear opening, you will be able to fill it from your side without fighting off hungry goats.

Floors and Bedding

Dirt flooring in a goat barn with a thick layer of bedding and litter material to absorb smell and moisture is adequate. Straw and waste hay are easy and economical, but wood shavings are easier for cleanups and more beneficial as future compost.

Goats will choose an area of their barn for sleeping unless you provide an elevated platform against a wall. They won't naturally keep this area clean, but it will stay cleaner than the rest of their living space if placed in the corner or in a low traffic area.

All sleeping areas should have thick bedding that is changed or added to regularly. Lay fresh bedding over the existing every few days and completely replace bedding material when soiled, damp, or smelling like ammonia. If at any time you can smell ammonia in the goat barn, make deep litter removal your top priority.

Waste Removal and Garden Compost

The rich organic waste material removed from the barn makes great garden compost. Goat droppings and litter will not burn your plants or delicate root systems if you were to use it right away on flower beds, but be sure to compost it for a six month minimum before using on vegetable gardens. If you've been using waste hay as bedding material or litter, I suggest composting it under a dark tarp or bin for at least a year or you'll have a miniature hay field sprouting in your gardens from all the seeds.

Lighting Increases Production

My first dairy goat shed didn't have lighting, and as the days grew shorter, I found myself milking and feeding in the dark more often than not. One simple light made all the difference to my comfort level, and it also kept the does in top production. Combined natural and artificial lighting for eighteen to twenty hours per day maintains milk production through the fall and winter months. It also increases the chance of an early spring estrous.

Running Water Saves Many Steps

If it is a special treat to have a water tap in the barn; it is ten times so to have both hot and cold water there.

If you are keeping dairy goats, you will save countless steps carrying milking paraphernalia back and forth to the house twice per day. Those trips become even more annoying when you or your doe have knocked over or dirtied one of your cups or buckets and you have to return to the house, resterilize everything and start over.

Hot and cold running water in the barn isn't a treat once you've had it for a few months, it's a necessity. You can properly wash up in the barn. You can start the cooldown process immediately for raw milk. You can give your goats the slightly warm water they love to have in cooler temperatures.

Safe Grain Storage

The idea of a goat breaking into the grain or feed bin, suffering from bloat, and dying before I have a chance to help is one of my biggest fears. It happens more often than you'd think, which is why I now store our livestock's grain in the basement of the house. It is convenient—the basement door opens to the back of the house, which is where I begin my chore journeys each morning and evening anyway.

Most people store their grain directly in the barn so it is on hand for on-the-spot ration adjustments or refills if one of the containers get knocked over.

Grain should be kept away from all moisture, out of the sun, off the ground if it is bagged, and definitely out of a goat's reach. A galvanized trash can with a snap-on lid placed well out of a goat's reach is perfect. They are cheap to purchase, last for decades, and also keep vermin out of your grain.

7

Pasture, Fields, and Forests

Yards or Acres of Sunshine and Fresh Air

TO PREVENT BACTERIAL infections a goat's hooves, should be on a dry surface for most of the day. If you are not pasturing goats and don't have a dry area for them to spend their outdoor time each day, a poured concrete pad for them to lie and walk on will suffice. At least part of that yard should be on the south side of the building where the sun's rays and warm breeze will dry off the land and give the goats a warm corner to enjoy the day.

RIGHT: A young Boer kid contemplates the ramp. Green pastures await below, but the sunny deck is warm and momma is still inside sleeping.

A fresh new field of pasture and all she can think about right now is testing that fence! While specialized nutrition and care are important, there is nothing that will keep your goat safer than a sturdy fence. It is back to the drawing board on this one.

Raised shelters are another option to keep goats high and dry during the rainy season. Commonly seen with a deck for goats to walk and lie on, these shelters have sturdy ramps that provide easy access to pasture.

Goats are happiest when they have something to climb on. An outcropping of rocks, a stone fence along the pasture, or a multilevel wooden platform will satisfy their instinctual nature and provide exercise for all their muscles. Keep climbing objects well away from fences though, or they'll use it as a step to freedom.

Fencing for Pasture or Yard

"A fence that can't hold water, won't hold a goat" is a truth-bearing, age-old axiom.

Above all other considerations, your fencing and gates deserve the most attention. Goats will go over, under, or through a fence before you've taken three steps away from their containment area if it hasn't been built correctly.

There are as many ways to build a fence as there are goats to climb over or crawl through them. Choose a method and materials that you are most comfortable with using my top goat-fencing tips.

- Four feet high minimum. Five feet high for highly active and nimble Nubians.
- Horned goats get caught in the openings of page wire fences but if you prefer this option and know that you will only keep polled goats, page wire fencing with twelve-inch inch openings is acceptable. If you will ever need to contain kids, invest in a six-inch page wire.
- Chain-link or stock panels with small openings work well to contain goats but are expensive for large areas.
- If you already have page wire fencing installed run an electric wire, nine to ten inches from the existing fence and about twelve inches off the ground. This keeps horned goats from

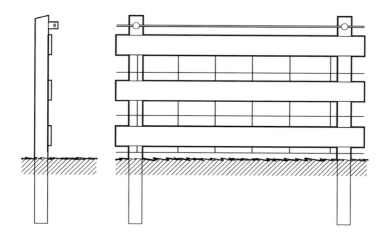

getting their heads caught in the page wire and ensures that the goats don't use the rungs of the wire as an escape ladder.

- Posts should be eight-foot high and buried at least thirty inches into the ground. Space them eight- to ten-feet apart. Posts can be steel or wood.

- Corner post supports go on the outside of a goat containment system. Goats will climb on, or shimmy up, a fence support no matter how slim the supports are.

- If building an electric fence only, run the wire, from the bottom up, at: five feet, ten feet, eighteen feet, twenty-four feet, and a final strand at forty feet for dairy, or forty-two feet for meat breeds. Electric fencing is not a viable option if hydro in your area is prone to blackouts.

The Gate

Your goats will watch you enter and leave the yard. In doing so, they will learn how to operate the lock. As soon as they've mastered the latch or handle, they'll be wandering through your flower garden, investigating activity on the road, taking their lunch in the grain fields, or bleating at your front door.

A goat can flip a hook out of the eye it rests in and has the determination to mouth and hoof at a lever latch all day until it opens. Determined goats have even been known to slide a large bolt to the open position.

- Place all slide bolts, latches, or locks on the outside of the gate where the goat can't reach them.
- Install your gate to swing into the goat yard so that even if one of your escape artists managed to unlatch the gate, she might not know she did so.

The Pasture

The practice of pasturing goats is a personal decision that may be based on breed of goat, farm economics, available pasture, or your need to have brush cleared.

As feed can be 70 percent of the cost of keeping any goat, fair weather pasturing will save you a lot of money.

Dairy does set out on a managed pasture of grasses will create more milk but it will be lower in butterfat content.

Pasturing dairy goats requires a little more care than setting meat goats out to browse. Dairy does that consume pungent plants will have off-flavored milk. You also wouldn't want to set your dairy doe out on pastures loaded with thickets and brush as her udder could easily be scratched or damaged while foraging in overgrowth.

Every farm is different, and every field will have a different rate of regeneration. The basic rule of thumb is to allow at least one acre for every ten goats and to employ rotational pasturing. As soon as one area looks sparse or the goats are spending more time eating hay at the shelter than time in the field, move them onto a different section of your land. This rotating ensures that each pasture area will remain viable, decrease the potential for parasitic infestation, and allow the goats the opportunity to browse on plants at their peak nutritional value.

Goats will eat a wide range of native plants on acreage but should still have access to free choice hay so that they are not forced to eat less than desirable forage. Your goat may have an instinctual nature not to ingest harmful plants, but you should never count on it. Take precautions by walking your pasture and knowing the plants growing there. Local authorities maintain lists of known poisonous plants in your region if you are unsure.

Even nonpoisonous plants can be toxic if they've been sprayed with pesticides. Fields that butt up to the road are a potential hazard for herbicide residue.

Keeping Goats Safe From Predators

It is a fact of life. Goats on pasture may be stalked, attacked, and killed by predators. It can happen in the daylight hours, but most often it happens in the evening when animals are resting peacefully.

Losing a prized goat or kid to coyotes, feral dogs, wolves, cougars, bears or wild cats is heartbreaking. You run through a wide range of emotions: sad, mad, sickened, and if it happens more than once, utter frustration. If you found your goat's remains on the field or in the nearby forest, the image will burn itself into your memory.

Every region across North America has cycles of predator population exceeding available wild food sources. When natural food sources dwindle, domestic livestock such as goats and sheep are the next available option for opportunistic wildlife.

Although our coyote population is growing, and three nights every week we hear them calling in the woods that border our fields, I have never lost a goat or kid to predators. I am not a stranger to predators attacking my animals though. Having lost countless ducks to

BELOW: A kid won't have a chance once a hungry coyote or the neighbor's dog enters the pasture. As their keepers, we have to take every precaution possible to keep our goats safe.

foxes, baby chicks to a raccoon, and one pup to a pack of coyotes, I understand the tragedy all too well.

Neighbors just a few miles down the road from me have lost more than a few kids. Neighbors just a few fields over have had sheep ripped apart on consecutive nights even though they have employed the use of a donkey, herd protecting dogs, and a llama—all standard and viable options to protect your herd.

If you have a problem with predators, you could try stronger electric fencing, which won't do much good if you are in an area known for having elk, moose, or deer charging through and pulling down fences regularly. Some farmers lay a grounded barbwire around the entire perimeter of pasture, which sends a strong message to wolves and coyotes stalking from the fence line.

The only protection I provide my goats is to lock them into the barn every night. There may be other reasons I haven't lost a goat to our local predators though. I keep my horses out at night, which may be too much animal for a predator to fight for a meal, and I am active on my land. Every week I walk the perimeter with a male dog marking his territory every five hundred yards. Finally, nearly all of our fields are visible from the house.

I think that given all these small annoyances to a wolf or pack of coyotes, my goats are just more trouble than the ones down the road. Hungry predators are lazy—they'll always go for the easiest meal.

No two situations are alike in the most effective legal manner to cope with these life-threatening pests. Trapping or hunting the predators are extreme (yet satisfying) options, but they may also be illegal in your area. Always check local regulations before taking matters into your own hands. Even though farmers have rights to protect their livestock, your problem predator might be a protected species.

Livestock Guardian Dogs

Guard dogs with a natural instinct to protect livestock are a perfect compliment to a herd of goats that are on pasture and out of your realm of protection. An entire class of dogs specific to the task have been bred for centuries in European countries.

LEFT The Anatolian Shepherd is considered to be the most intelligent of the Livestock Guardian Dog breeds and have been used to protect family members, sheep, miniature horses, and goats to name a few.

Livestock Guardian Dogs (LGD) have been known to protect goats, alpacas, sheep, and miniature horses from bears, coyotes, feral or domestic dogs, and wild cats. With an overdeveloped sense of smell and night vision, there is little chance that any predator can approach them.

Commonly found purebreds that fit into this class are the Anatolian Shepherd, the Great Pyrenees, and the Komondor. If you are considering this option, you may also run across the Akbash, Maremma, Karst, Ovcharka, Tatra Sheepdog, Shar Planinetz, Slovac Cuvac, and Kuvasz in your research. Get to know all their breed strengths and differences before you make a decision on which guardian dog to pursue, and take even more time to find the best of the breed.

Although most of these dogs are highly intelligent and protective, you will find some older dogs being offered at an incredibly low price. These dogs may have been raised as a pet instead of protector (which renders them pretty useless when you introduce them to your herd) or may not be physically equipped to handle your climate. There is no finer place to buy an LGD than directly from the breeder who can teach you many of the finer points in training and what to expect in performance from every one of his pups.

If you have other animals on your property such as cats, chickens, or horses, you will want to get an LGD puppy to socialize him to the other animals at a young age. An unsocialized older dog may see the other animals as a threat to his own herd. LGDs are not ready to protect a herd until they are eighteen months old, so if you need protection right now, ask the breeder for suggestions to keeping your herd safe until your pup has matured.

Natural Herd Protectors

I would be remiss to make such a fuss about the Livestock Guardian breeds without some introduction of other animals used for protection of goats, namely the donkey and the llama.

Similar to guard dogs, donkeys have a natural instinct to protect their herd mates. They are ferocious in their attack and will fight until the bitter end even when another animal is not under attack but being stalked.

I once watched a pack donkey at the back of a lineup keep a watchful eye on a large cougar for over a mile. Although the cougar never attacked, he either came one inch closer than the donkey could tolerate or the donkey just tired of having the wild animal following him and he charged. Within five minutes, the vibrant cougar lay motionless and still forevermore. The donkey hadn't even suffered a scratch.

Llamas can be equally ferocious in their attack, but they are best suited to warding off canine predators such as wolves, coyotes, and dogs. There have been reports of llamas, knowing their limitations, running to the farmhouse to alert their human keepers of trouble within the herd when mountain lions or bears were stalking the goats.

Taking on a new animal to protect the herd will require knowledge to keep that animal in good form. Llamas and donkeys already have a instinctual protective quality, and in most cases, no training will be required.

LEFT Guard llamas also defend against predators by making an alarm call, leading the herd to safety, or by chasing an intruder away.

⑧ Feeding Goats

THE LARGEST MISCONCEPTION in livestock is that a goat will eat everything in its path. They are depicted as chewing on tin cans, eating the tires off a bicycle, and annihilating the flower garden faster than a few hungry gophers. Having lived without good fencing for many months, I say that the goat gets a bad rap. Our horse chewed and demolished ten items to every one treasure that fell prey to the goats' tasteful exploration.

Exploration? Yes, because that is all it is for goats. They are not voracious empty pits that will eat everything in sight should they escape. They're oral explorers, eager to mouth all new objects that grace their line of sight. At times, in order to fully understand an item, they may rip at, tear apart, and shove your treasured possessions, but we can't fault a goat for inquisitiveness. What we need are better fences!

For their safety and that of your personal property, design a great fence with the understanding that one day they will challenge it and escape into your rose garden. Reason enough I'd say to reinforce the corner posts and add a solar inverter to the electric wire.

BELOW: Contrary to popular belief, a goat won't eat everything in sight; they're just curious about the world. Without fingers to feel the texture of a denim shirt; their mouth is their only venue to discovery.

When I finally wised up about fencing, I moved my goats into the overgrown fields, supplied plenty of fresh water daily, and left them to live off the land for three seasons. Thankfully, those fields supported them without any deficiencies. (It is important to note that these were old grain and hay fields with varied forage.)

As winter approached, I began shopping around for hay to feed them once the land would no longer support them. This is when I gained a clearer picture of nutritional requirements, variances in hay content and quality, plus the required supplements. I found a feeding program that worked for our farm style and budget as well as good goat nutrition, which was far simpler than the mass of calculations and conditions that are so graciously published by the complete feed and nutrition companies.

When I sat down to write this book, I surmised that you should have a program that was more advanced than my own. That for the health of your goats, you should have all the information at your fingertips for precise nutrition. I researched for days on vitamins, nutrients, minerals, and percentages of proteins, to energy, to fiber. I then contemplated the needs of a lactating dairy doe versus a 320-pound meat breed wether and wondered aloud how anyone could ever stay on top of all that.

My final conclusion is that we don't need to match the precise requirements to the nth of a milligram to keep our goats healthy. I say this with exactly the same vigor that I employ when considering my own health and nutrition. Yes, I am aware that frozen vegetables don't deliver anywhere near as many nutrients as fresh from the garden, but in the winter we eat them anyway.

My point in sharing all of this with you is to ease your conscience about not following some precise list of food requirements in exact amounts. If you provide variety to a goat and don't force them to eat a particular food out of starvation, they will take what they need from that variety, and their good health will (more often than not) carry them through the winter months.

I do understand the need to manage and control what goes into the animals that eventually go into us, but I am always amazed at the people who get more involved in their animals' nutritional requirements than their own. In the end, you'll do well to exercise common sense, learn about the basics that you'll find in this chapter, have a good idea of the quality of forage on your land or in the hay you buy, and research the rest on a need to know basis. The Internet is a valuable

LEFT: Goats love to climb. It is part of their need to explore the world from every angle possible.

tool for specialized knowledge, and your veterinarian should be your greatest ally.

Variety In Their Diet

Goats may be one of the hardiest of farm animals, but their digestive system and nutritional needs beg for variety in their diet. In the wild, a goat will wander from one area to the next, grazing in green fields as often as browsing through overgrown or wooded areas. They will eat leaves, branches, bushes, brush, brambles, vines, grass, weeds, and tree bark. Most goats have an instinctual ability to stay away from poisonous plants or at least not partake in them when they are at the height of toxicity.

It is the goats' ability to utilize such a wide variety of food sources that makes them an easy keep. By choosing to keep goats, you will have the opportunity to set them out into pasture, allow them to clear land of brush and brambles, or purchase an all-encompassing feed for them while keeping them in a barn with an exercise yard.

Feeding Styles Based on Goat Type

As feed will account for the majority of ongoing costs associated with keeping goats, allowing them to pasture or browse will lighten the financial burden considerably as long as it suits your purpose.

BELOW: Goats are a relatively easy keep. You can set them on pasture, in overgrown fields, and even partial forests. As long as there's enough to eat, they'll thrive.

Just as there is a best breed for your intended use, so will there be a best method for keeping and feeding that breed. A dairy doe, for example, would not be set out into overgrown fields of raspberry brambles and spiked saplings as she could damage her udder. Furthermore, some wild plants will unpleasantly alter the taste of her milk. You would be better served to keep that doe on a field of managed pasture full of nutritious grasses.

As with every rule to keeping farm animals, there is an exception to the pasture-for-dairy-goats rule. If you plan to make cheese year-round, a confinement system of shed and yard might be the better option as does fed mainly on lush green pasture will produce a thin, lower butterfat milk.

Goats raised for meat or fiber are the most versatile on the types of terrain they can find sustenance on. The one exception is the Angora breed. Mohair from the Angora goat will net higher returns when it is free of twigs, leaves, and bugs picked up from time spent in fields or forest.

BELOW: Brush and bushes like this thorny weed are soon banished from your fields. Although the goats may not eat the entire plant in one pass, eventually, the weeds that decrease the value of your fields will just stop growing.

The Components of Nutrition

Your goats require adequately nutritious foodstuffs to encourage growth and production and support good health. The main components of nutrition are energy, protein, fiber, vitamins, minerals, and water. Deficiencies, excesses, or imbalances of these can lead to a variety of health problems—not just disease or susceptibility to infection.

Fiber will be the main source of your goat's intake, mainly supplied through grasses, browse, and hay. Although hay is nothing more than pasture plants that have been cut and dried, all hay is not created equal. Much like vegetables for human consumption, different field grasses supply different vitamins and nutrients for the goat. Increased roughage during cold weather helps goats to maintain their body temperature as rumen activity requires and produces energy.

In the simplest diet, you would allow goats plenty of fiber, some grain, and a vitamin-nutrient supplement. You won't wander too far away from good nutrition by following this simple plan as long as you continue to supply hay for variety when goats are on pasture. As for grain, you will be making ongoing adjustments to quantity based on your goat's body condition (i.e., fat, thin, breeding, pregnant, lactating, growing) and type.

Although you can study, analyze, test, and assess the quality of your own pasture or the hay you buy—and doing so is honorable—it is often unnecessary for small farms. Mineral blocks created to supplement goats fed mainly with hay will make up for many deficiencies. Should you opt instead for a concentrate or ration created for goats, you'll find all required vitamins and nutrients within the bag.

The outcome of my lackadaisical warm-weather feeding plan from the start of this chapter would have been different if my fields had been full of lush pasture instead of varied grasses, shrubs, weeds, and grains. A goat can starve on a solely pasture diet. Growing

ABOVE: Goats like variety in their diet. They'll look for roughage and fiber high and low and are better than most humans at knowing what they need more of in their diets.

grasses have such a high volume of water that a goat cannot eat enough during daylight hours to be satiated. The trouble doesn't end there. An abundance of fresh pasture in a goat's digestive system produces excessive gas, which could lead to the physical affliction of bloat, often resulting in death. When you've put your goats on pasture, be sure to supplement with free choice hay and grain daily for the dairy does.

If you're pasturing during the warmer months and your goats are bedding down for the night behind closed doors, a ready supply of hay throughout the day and night will alleviate the chance of bloat. They will eat hay in the morning while they wait for you to let them out and return to it intermittently throughout the day. This balance ensures that they have enough food for their bellies, fiber-rich roughage even if they're on grassy pasture and enough variety between the two.

The Current State of Commercial Feed

With every year that passes, I grow increasingly concerned over modern advances in animal husbandry. In my research for this chapter, I discovered a well cited online tool for calculating the nutritional needs of every classification of goat at varying stages of their lives.

Once all the questions had been answered about the goat in question, you could then select from a wide variety of food sources. The calculation helps you determine if you are aptly meeting the animal's dietary requirements. Within the final selection of ninety-nine food choices were three that stopped me in my tracks: dried poultry litter, dried poultry manure, and hydrolyzed feather meal.

I admit that I am an idealist by nature and that the documentary and supporting research behind Food, Inc. has changed the way I observe commercially raised meat, but after keeping farm animals for many years, I know that the closest you can feed an animal to what their ancestors ate in the wild, the better. No self-respecting healthy goat would ever choose to eat these three—as well as a few of the other items in that extensive list. Sure, grind it up, cook it, and smother it with molasses—they *might* eat it, but to what end?

Understanding the Goat's Digestive System

Digestion is the process of breaking down feed material into components that can be absorbed for nutrition. It is a goat's amazing digestive system that allows them to eat and utilize the widest range of plant matter than any other farm animal.

A goat's digestive system takes up a third of their body and contains many parts that need to function together perfectly to ensure optimum health and production. The most amazing feature of a goat's digestive system is the rumen (the first of four stomachs).

This first stomach is akin to a three- to five-gallon fermentation vat. Muscular contractions within the rumen break down plant fiber and begin to manufacture many important vitamins. At the same time hay, grain, and water blend with naturally occurring bacteria and protozoa. The bacteria and protozoa secrete enzymes to facilitate a chemical breakdown of food (fermentation).

In the rumen, carbohydrates are broken down and converted into simple acid molecules absorbed directly from the rumen's walls into the bloodstream. These molecules will either be used for immediate energy or stored as fat. Proteins are broken down into amino acids and then into fatty acids before conversion is made to a usable protein source.

A goat at rest will regurgitate food from the rumen, chewing it again thirty to forty times before swallowing. Known as "chewing their cud," that regurgitated food is then passed into the reticulum (second stomach), the omasum (third stomach), and finally to the abomasum (fourth stomach). From the second to the fourth stomachs, the food is ground smaller with each passing until it eventually moves into the small and large intestines. Within the intestines the majority of nutrients will be carried into the bloodstream for immediate access.

BELOW: This stunning little doe is taking a rest from eating to enjoy the warm day and regurgitate her cud.

Goats love grain! These two kids have already finished their small handful of grain but that doesn't mean they can't look inside and hope for more.

Rumen Health and Bloat

With the tall order the rumen fills, you can see why optimum performance of this organ is vital. Other than ensuring your goat gets the best feed in the right quantities, it seems there is little we can do to manage this aspect of her health. Knowing and watching for external clues of healthy rumen activity is a proactive step. All too often, once we notice that rumen activity has failed, it is too late; results could be fatal.

Goats with a healthy rumen exhibit two easily recognizable signs: a persistent and audible gurgling digestion (when you put your ear to their abdomens), and belching every twenty to thirty minutes. Belching releases the formation of gas created during normal digestion.

The most common complication of inadequate rumen function is the condition of bloat. Unreleased gas builds in the first stomach and blocks air passage to the lungs. The result is death by suffocation in under sixty minutes. Lack of balance is the culprit—either too much grain without fiber or too much green grass without fiber. Without fiber to solicit cud regurgitation, foodstuffs ferment and the resulting gas becomes trapped.

Here is a true lesson in how to kill an animal with kindness. Goats love grain, and we enjoy spoiling our animals, but too much will certainly kill them. Although I have personally never lost a goat to bloat, I have heard my share of sad stories on how this happens—from well-meaning chore hands, "a little extra to get you through the night," and unsecured grain bins. Lock up your grain bins and educate every member of your family and well-meaning neighbor. Every person who ever comes into contact with your goats needs to understand the devastating effects of overfeeding. It could just save a life.

Should you suspect that your goat has broken into the grain bin and gorged, you should immediately call your veterinarian out to the farm. While you wait, watch over her carefully and closely monitor rumen activity. Depending on how much she's eaten, a condition of bloat could develop but not escalate to life-threatening. Your vet may suggest isolating her and feeding her a good mixed hay in small amounts repeatedly through the next twenty-four hours. Once rumen activity has returned to normal (belching and gurgling sounds heard in the abdomen), she can be returned to the herd with a normal feeding schedule.

Soda

Goats need to maintain a low level of acidity in their four-part stomach. Grain and rich pasture may increase acidity and upset their delicate balance. As you've read in the previous section on Rumen Health, increased levels of fermentation within the rumen could prove fatal.

Given the correct supplementary aids, a goat will self-manage their acidic range when on a normal diet. Access to soda is often all that is required.

Recent studies in goat health suggest that soda is harmful to wethers, especially if they were castrated younger than three months of age. Please consult with your veterinarian if you are in doubt about the use of soda, the age at which your wether was castrated, or if you are concerned about your goat's ability to self-manage acidity levels.

The average dairy goat consumes up to two tablespoons of soda per day. Feed-grade baking soda is available at your feed store, but the grocery store variety will work without harming them; it just costs a little more.

Factors Affecting Feed Consumption and Quality

A goat will consume 3 to 10 percent of their body weight daily with variances given to breed, weight, physical condition, and age. Three other major contributors to dietary needs are the demands of pregnancy (or in a buck's case if he is or will be breeding), lactation, and hair growth (fiber production goats).

External factors affecting feed quantity are those of activity and climate conditions. Goats grazing on physically difficult terrain require more feed to balance the extra expenditure of energy. Keeping a healthy body temperature in colder weather is another drain of their energy resources. More hay will be also be consumed when it rains, with the minor exception of the meat breeds venturing out to graze and browse in light summer rains.

Gestation and lactation are two critical periods for a doe's nutrition. Never discount the demands of pregnancy or lactation as a job they were created to perform. A fetus-carrying or milk-producing doe's health will be compromised if she is not given ample nutrients, minerals, water, and feed.

Kids also require extra attention. Kids fed incorrectly may never reach their genetic potential. Start kids off with the right nutrition, and they will reach their full potential as adults. If you are only growing kids to a market weight in the shortest amount of time, you'll soon discover that goats kids will not fatten like cattle. Free choice grain, fed in an effort to increase weight, accumulates fat, but it will be mostly around their internal organs. Goat kids require the right foodstuffs for their age and breed to grow. We cannot simply supply them with any food and expect them to grow at our preferred rate. Meat-breed kids make good use of extra protein-rich feeds as it helps them to gain healthy muscle instead of pockets of internal fat.

Common Complications of Incorrect Nutrition

1. Abortion
2. Acidosis / Indigestion
3. Arthritis
4. Diarrhea / Constipation
5. Ketosis
6. Lameness
7. Parturient Paresis (milk fever)
8. Poor Colostral Quality / Quantity
9. Starvation
10. Urinary Calculi
11. Vitamin / Mineral Imbalance
12. Weak Kids at Delivery

Winter Feeding

As you've read the best way to feed a goat is to offer variety—just as a goat might find in the wild. I am not proposing that you harvest sticks, bark, and leaves in the middle of January, but that through offering a variety of nutrients, vitamins, and fiber through hay and grain (or a fully balanced concentrate), your goats will winter over nicely.

This is the season where the difference between dairy and meat breeds nutritional requirements are noticeably different. With wintered-over meat breeds, we focus more on their protein intake than their feed variety.

You'll read below that all hay is not created equal and that it is impossible to tell the protein content of hay without laboratory analysis. On average, grass hay has a protein content of 8–13 percent, which puts it in perfect range of a meat breeds' protein needs. However, the meat breeds are commonly fed a variety of protein-based foodstuffs as well as higher protein hay to compensate for lack of variety in forage. You will find the protein content of various food enhancements

Dairy goats are
often set on
gentle fields of
lush pasture, but
they need variety
to round out their
diets and keep
them healthy.

as well as the requirements of various stages of development for all breeds at the end of this chapter.

To provide variety to dairy goats in the winter months, you could offer some chopped produce. Minute amounts of green beans, turnips, pumpkins, beets, cabbages, and carrots, or a chopped apple or pear are favorites in our barn. Provide these in moderation; variety is the path to good health.

As with any animal, nutrition affects not only the overall well-being of a goat but also milk production, reproductive success, growth, and physical condition. Goat experts agree that more than 60 percent of total production costs—whether used for meat, milk, or fiber—is directly attributable to meeting the daily nutritional needs of each animal. It certainly pays to pay attention.

What You Need to Know About Pasturing

Green pasture, forbs, weeds, and browse are the most economical source of energy and nutrients for goats. In most cases, meat breeds will thrive in overgrown fields and terrain less palatable to other livestock. This makes them excellent companion animals to cattle and sheep.

Goats won't be competing for the same food source as a primarily grazing animal. As with all things livestock, there are a few caveats to this concept.

First of all you need to know your land and what it has been used for. Many areas of North America have soil that is depleted in minerals and nutrients. Selenium depletion is common; plants growing on depleted soil will not completely meet your goats' requirements. Personal communication with your veterinarian, local goat owners, and the farm supply store staff will provide clues to required supplementation for your area and your breed choice.

Secondly you will want to watch the growth and regeneration of pasture lands. When pasture is green and growing, it is very high in energy, nutrients, and protein. If particularly lush, it can also have a high moisture content. Goats consuming an abundance of moisture-rich feed will need to eat more to meet nutrient requirements.

Grass tetany (a metabolic imbalance of magnesium) and bloat are two conditions that can occur in goats of any stage of life when given lush or leafy feed with too radical of a change. Take special care in the spring when grass is growing the fastest, on pasture that was purposefully seeded in past seasons, and with lactating does.

If you have ample land or if you are keeping goats to control brush on your land, consider rotational grazing of smaller pastures. This ensures that plants at various stages of moisture and nutrition are eaten and digested in a balanced manner. Rotating pastures ensures the plants within are kept in a vegetative state and therefore are more nutrient rich.

The Ins and Outs of Goat Hay

Hay Testing

The only way to truly know nutritional content of any hay is to have it analyzed in a laboratory. Although this may appear to be overkill, many large herd owners and dairy farms do utilize the services of a forage testing laboratory for analysis. In the United States, samples can be brought to a local county Extension office. Expect to pay between $10 and $20 for analysis with an average two-week turnaround.

Storing hay on your own property as soon as it comes off the farmer's field is both economical and wise. You don't want to be scrambling for hay in February, paying a higher price per bale, or worrying about moving it from one barn to your own during freezing weather.

As long as your goats are healthy, they won't get fat on hay or eat more than they require. You'll note that I keep free choice hay available at all times of the year. When pasture is sparse or when grazing is impossible due to rain or snow, hay will be your goats' primary nutrition source.

All hay is not created equal and the debate over which hay is best for goats is a heated one amongst many owners and experts.

Hay that is adequate for cows will not provide enough nutrients for goats. Cow hay is just barely suitable as bedding for a goat. You need hay fit for a horse or better. A 50:50 ratio of legume to grass is perfect for most goats.

For our farm, keeping a variety of hay on hand has worked well. We serve up a good-quality, mixed pasture hay that has been grown for horses and throw in a few handfuls of legume hay for the dairy goats. Legume hay—harvested from fields cross-planted with a mix of alfalfa, clover, soybean, vetch, or lespedeza—tends to be higher in protein, vitamins, and minerals than grass hays. Kids and pregnant or lactating does will benefit from the extra nutrition that legume hay provides.

The energy and protein content of any hay will rely on the maturity of the plant when it was harvested.

Aside from harvesting time, proper curing and storage is also necessary to maintain nutritional quality. Hay should be kept high and dry, preferably out of harsh winds and direct sunlight.

- Hay that is a little dusty can be given a good shake before adding to a manger or feeder, but don't be surprised if your goats turn up their noses anyway. (Shake hay away from the goats to avoid unnecessary inhalation of fine dust particles.)
- Hay that has become brittle from age or sun damage has minimal nutritional benefit and is better suited for bedding.
- Hay that shows mold within the bale should be disposed of. It is unhealthy, and they won't eat it anyway. If they eat it because they have no other food source, they could get sick or develop respiratory problems.

It is completely normal for goats to eat the best part of the hay and leave half of the bale in the manger. Don't let old hay sit in the bottom of the manger. Rotate it for a day perhaps, then delegate it to their bedding.

How Much Hay?

The average dairy goat will eat 3 to 5 percent of her body weight in hay each day. The average meat goat eats five percent or more. The average square bale weighs 35 to 40 pounds. Therefore to keep a dairy goat in hay and without pasture, given an average weight of 120 pounds, you'll need approximately thirty-five to forty square bales of hay per year. Consequently to keep a meat goat in hay and without pasture, assuming a weight of 200 pounds, you'll need eighty to eighty-five square bales per year.

Ration, Grain, and Supplements

Much like a human, a goat's needs will change as they mature, as they carry their young, and as the seasons change. Whether your goats are pets or for production, you will want to be acutely aware of each goat's daily ration or grain intake and regularly make adjustments based on their current condition (which you can monitor through a body condition scoring assessment from chapter 1). Assessment and feed adjustment is an ongoing process.

LEFT: Although they may look chubby, the African Pygmy was bred to be barrel chested and stocky. Their current shape isn't completely due to too much grain and not enough exercise.

A variety of goat rations will be available at the feed store. Each should be labeled with counts for nutrients, vitamins, and protein. If your local store doesn't stock the specific feed you need, ask them to order some based on your criteria. A lactating dairy doe, for instance, requires 16 percent protein in her diet, while meat goats and wethers only require 12 percent.

The composition of feed has changed over the years and varies by region. Extra minerals and vitamins are often supplied through supplemental rations. A prime example is selenium. Recent research is also uncovering tocopherol (vitamin E) depletion. Feed manufacturers are aware of local source depletion and may add trace nutrients to their feed ingredients for the area they're serving.

Pay close attention to the labels and quantity suggestions and by all means collaborate with local herd owners, your veterinarian, or goat associations regarding supplemental nutrients not being met naturally in your area or specific to your goat's breed and body condition.

Assessing Goat Condition to Modify Consumption

From the moment your goat steps foot on your land and until the day she leaves, you will regularly assess her body composition and condition. From your assessment you will be altering grain consumption to increase or decrease weight, to flush eggs from a doe's reproductive organs, and to ensure that every goat is in optimum health.

Does of any breed should not be allowed to stray from their best weight. If too thin and lactating, they may be using all of their energy to produce milk at a detriment to their own current and future health. They may be incapable of impregnation, have fewer kids, or not be able to look after meeting the nutritional requirements of the kids they have.

Overly fat does could have their own series of complications during labor and delivery (breech births and toxemia are the most common), but both sexes will suffer when overweight. A goat has a natural tendency to store fat in the abdomen first—surrounding the liver and kidneys—before showing up on the extremities. Fat stored around organs is especially dangerous to the health of any animal.

Assessing the condition of all goats is just one of the reasons you want to spend hands-on time with your animals every week. It won't take much practice to know the difference between a layer of fat versus muscle mass.

Should fat begin to accumulate, you will feel the difference first along the ribs and spine. Push gently in all other areas including shoulders, the tail head, pinbone area, and the edge of the loins.

The other extreme (too thin) is to feel bone directly under the skin of the goat along the backbone and ribs. Somewhere between too thin and feeling a thick layer of fat on the animal is your goal. Adjustments in grain, ration, or concentrates is the method of correction.

Remember to make all changes in diet gradually. I find that if I add an extra ounce of energy concentrate to the doe's grain bucket every day for four days, then feed her a full four ounces every day for a month, she'll stabilize at her best weight for the remainder of the season. The opposite—cutting back on grain a little at a time—works for fat goats.

Grain and Goat Ration

Most goats will require grain at some point for health, energy, and weight management. Goats in production require even more. Nutritionally designed goat ration (also called goat chow) exists to match a goat's age, breed, purpose, and current condition.

The meat breeds are usually only fed grain or ration during winter months in the form of protein feeds (see chart at the end of the chapter) to supplement free choice hay. Almost all other goats will eat grain or ration every day with the exception being the overweight goat.

Use the chart at the end of the chapter as a starting point to determine your goat's needs and suggested quantity. Adjust quantities based on physical condition, seasonal changes, and the quality of hay you're feeding. With minor adjustments throughout the year, you will find the perfect quantity for every goat in your barn.

Complete Concentrates

Concentrates are another way to feed or supplement a goat. Concentrates may be a good solution if:

- you are feeding a low-quality hay or grazing pasture,
- your does have been bred or are lactating,
- your kids need more food and nutrients than they can get on pasture daily,
- you are nurturing a goat that is failing, or
- you have limited storage for hay.

To choose a good concentrate you need to know that there are three types available: energy based, protein based and complete nutrition. If your feed store is uncertain of the type it carries (or can order for you), you should be able to sort it out just by looking at the product ingredient list.

Energy feeds typically include cereal grains (corn, barley, wheat, oats, milo, rye). If you're limited to feeding your goats entirely by concentrate, you'll need to take extra precaution in how the manufacturer has balanced out the nutrients.

At the basic level, energy comes from carbohydrates (sugars, starch and fiber) and fats. The rumen's natural bacteria ferments carbohydrates and fats into fatty acids, and those acids are absorbed and used for energy. Fat is efficiently used for energy but should be kept below 5 percent of daily intake to ensure proper rumen activity and healthy conformation.

Typical cereal grains in energy feeds are generally high in phosphorus but low in calcium. All that phosphorus without ample calcium to offset it doesn't fare well for goats and could cause kidney stones in bucks and wethers or milk fever in does. Whenever anyone gives you advice about adding a calcium supplement to your feeding routine, you can bet they're feeding, and compensating for, a high-phosphorus feed. Check ingredient and composition labels and aim for two parts calcium for every one part phosphorus for lactating does and lower for bucks, wethers, and dry does.

Protein feeds are most often composed of soybean and cottonseed meal. Although not often used by small farms as a food source, they are recommended to round out and balance the nutritional requirements of the meat goat breeds, especially during winter months. Often offered in varying protein percentages, choose the one you need to provide correct protein requirements based on your meat breed goat's age and condition and the estimated protein value of their hay. (See both charts at the end of this chapter for protein requirements and estimated protein values in varying feeds.)

A little excess protein won't bother most goats—since microorganisms in the rumen manufacture their own body protein, excess is burned off as energy—and may even help goats build a stronger immunity to internal parasites. On the other side of the equation, a lack of sufficient protein adversely affects milk production, reproduction, growth rate, and resistance to disease.

Complete nutrition feeds provide a balanced diet for your goat in any stage of life or health. Always communicate with feed store staff, feed manufacturer representative, or company website before purchasing and using any complete nutrition feed, especially if this is all you will be providing. The age, breed, current condition, and intended production of your goats are important factors in the balance of nutrients, energy, fiber, and protein composition.

Vitamins and Minerals

Your goat needs an adequate amount of vitamins and minerals to keep her in top condition. Since pasture and forage may fall short during different seasons, mineral blocks or loose minerals that have been formulated specifically for goats are available at every feed store.

Vitamin A is converted from the carotene found in green, leafy feed. Excess vitamin A is stored in the liver for three to six months when intake exceeds requirements but the reserve should not be relied upon for later access. Supplementation is almost always required. Your veterinarian may also suggest a vitamin A injection.

Vitamin C is self-manufactured in healthy muscle tissue in adequate quantities.

All the B vitamins and vitamin K are self-manufactured in the goat's rumen.

Exposure to sunlight helps goats synthesize vitamin D. Good-quality hay that has been cured in the field on sunny days may also contain vitamin D.

Salt, calcium, phosphorus, selenium, tocopherol, copper, zinc, magnesium, potassium, sulfur, iron, iodine, manganese, cobalt, and molybdenum are required in their diet for health and fertility.

Supplements are offered to compensate for any lack in feed quality. This is similar to humans taking a daily vitamin. We try to eat only food that will keep us healthy; we hope to find balance in our meals, but vitamins geared specifically to our sex, activity level, and age are a little extra insurance that we aren't missing prime nutrients. For the most part, our bodies will only use what we need and the rest will be flushed through our systems without causing harm. Is this a perfect measurement or an exact science? Not all the time. Research labs and scientists are always working to improve methods that maintain health in humans, pets, and livestock.

Goats that are eating a complete nutrition concentrate might not need added mineral or vitamin supplements. It is suggested by feed manufacturers that all the necessary components to health are formulated directly into the pellets.

Goats on hay and pasture are supplemented in one of two ways. A loose mix or a mineral block set out as free choice. Loose mixes can also be measured and added to each goat's daily grain ration.

Neither method is superior. I prefer the block, but I have been told that some goats have an aversion to them. The lesser the quality of pasture plants or hay, the more a goat looks for supplementation. In those situations, a goat could get a sore tongue trying to get all the required nutrients from a block. If you try the block and your goats spend a lot of time at it, or don't use it at all, try leaving out a loose mix.

I have read about owners adding loose mix to a goat's daily grain bucket. I wouldn't trust myself to make an accurate-enough assessment on other foodstuffs to know how much to provide per feeding. If my measurement is repeatedly wrong, I could be compromising the health and digestive balance of the animal.

In my experience, goats grazing on pasture and hay know when, if, and how much they need. Ready access to a block or loose mix in a mounted bucket ensures that every goat gets the minerals and vitamins they need, when they need it. Pregnant or lactating does have the highest need for mineral supplementation.

Mineral blocks and mixes make up for all the shortcomings of the land, hay, and ration. I have never heard of a goat overdosing from a block formulated specifically for them. I have heard, however, of livestock becoming sick from access to the wrong minerals. This is a common problem for sheep having access to a goats' minerals.

Water: The Elixir of a Healthy Life

Clean fresh water, readily available, is critical. The average goat will consume between one and four gallons per day.

GOATS NEED MORE WATER:

- when temperatures rise above 70 degrees Fahrenheit or drop unexpectedly,
- with increased consumption of dry feed,
- with increased exercise or other energy expenditure,
- during gestation and lactation,
- during growth spurts.

GOATS MAY NEED LESS WATER:

- when on lush pasture or moisture-rich forage,
- if pasture is covered with dew or snow,
- if exercise is restricted.

A lack of water, sullied water, or dirty water buckets can result in health problems. Some are fatal.

Water buckets must be kept clean. Watering bowls or buckets need to be emptied and rinsed daily and refilled as many times as necessary throughout the day. Sanitize all buckets or automated water systems weekly to prevent bacteria-related illness. I give our buckets a quick light spray of household bleach and then rinse thoroughly inside and out until the smell of bleach is gone.

The best way to supply water to goats is by head access only. Place water buckets outside of their pens where they cannot spill or soil them, but where can reach their heads through to drink.

Goats prefer lukewarm water, so provide it whenever you can. It is believed that feed quantity and nutrient requirements increase when goats only have cold water to drink. Although I haven't found any supporting research on the topic, I have almost always given our goats warm water during evening chores. Not knowing whether or not it made a difference in feed costs, it just seemed a small act of kindness to animals bedding down for the night in a cold barn.

> If the water you draw is from a well and it is hard—leaving deposits in your sink or buckets—it might be worthwhile to have it tested, especially for high levels of calcium or phosphorous. If found, you can filter the water at the barn or adjust the amounts given in dry mix mineral supplements.

GENERAL RATION GUIDELINES

Type	Condition	Daily Amount (in pounds)
Kid	Nursing	If interested, a bite or two
	Weaned on pasture	1 / 4—1 / 2
	Weaned no pasture	1
Bucks, Wethers, Open Dry Does		1 / 2 (maintenance)
NonDairy Does	Bred and Dry	1 / 2—1 (maintenance)
	6 weeks before kidding	1 (concentrate)
	Nursing	1—1 1 / 4 (concentrate)
	12 weeks after kidding	1 / 2—1 (maintenance)
Dairy Does	Bred and Dry	1 (concentrate)
	2 weeks before kidding	Up to 3 (concentrate)
	Lactating	1 plus 1 / 2 per pound of milk produced (concentrate)

Some quick adjustment guidelines per individual goats are:

- decrease ration to goats on pasture, overweight goats, wethers, and dry (nonlactating) does.
- increase ration to recently weaned kids, underweight goats, goats on pasture in bad weather, and pregnant or lactating does.

SUGGESTED MEAT GOAT BREED RATIONS

	Recommended Protein %	Ration Supplementing 11 % Grass Hay
Weanling	14	1 pound of 16% protein feed
Yearling	12	1 pound of 14% protein feed
Bucks	10	1 pound of 16% protein feed
Dry Does	12	1 pound of 14% protein feed
		Ration Supplementing 13 % Mixed Hay
Late Gestation Does	11	1 / 2 to 1 pound of corn
Average Lactation Does	13	1 / 2 to 1 pound of corn
High Lactation Does	14	1 / 2 to 1 pound of corn

PROTEIN CONTENT OF FOOD SOURCES FOR GOATS

Source	Protein %	Source	Protein %
Acorns (fresh)	5	Honeysuckle (mature)	10
Alfalfa Hay	13–20	Juniper Leaves	6
Barley Grain	13.5	Kudzu (early)	16
Bermuda Hay (7 weeks)	9–11%	Mimosa Leaves	21
Bermuda Hay (12 weeks)	7–9%	Mulberry Leaves	17
Chicory	15	Oak (buds and leaves)	18
Clover Pasture	25	Oats	13
Corn Grain	10	Pasture (vegetative)	12–24%
Corn	9	Pasture (mature)	8–10%
Cottonseed	22	Pasture (dead)	5–7%
Curled Dock	13	Persimmon Leaves	12
Fescue (6 weeks)	8–11%	Shelled Corn	11
Fescue (9 weeks)	7–9%	Soybean Meal	44
Hackberry (mature)	15	Soybean Hulls (ground)	14
Hay (Grass)	11	Soybeans (roasted)	43
Hay (Mixed)	15	Sumac (early)	14
Hay (Legume)	18	Sunflower Seeds	18
Honeysuckle (leaves and buds)	15	Wheat Middlings	19

⑨
Goat Health

THERE IS JUST no way around it; the opportunity for an introduction of illness, parasites, toxic reactions, and bacterial infections are everywhere on a farm. At times the cause will be within your preventative power, at other times the cause will be out of your control.

Being overly protective of your goats is not a character flaw. While those without goats may scoff at the idea, compassionate keepers will feel justified in their sensitivity, and the logically minded will recognize this as simply taking care of what is yours. No matter how you justify attentive care, we are their keeper and their health is our responsibility.

Here are my top 10 tips to keeping healthy goats.

1. Only buy from responsible sellers.
2. Only buy healthy animals.
3. Quarantine new goats for at least thirty days.
4. Strive towards cleanliness in the barn. You can't (and shouldn't) sanitize daily or pick up every piece of waste matter as it falls, but ensuring that fresh water is always available and bedding is dry and clean are two easy tasks that simply can't be neglected with goats.
5. Create and adhere to a vaccination schedule on your veterinarian's advice.
6. Be regimented in your schedule. Goats are creatures of habit and lovers of schedules—especially the dairy breeds. Lactating does know when the milking hour draws near, and they'll often be waiting at the barn door twenty minutes before your regular time. If you provide daily rations of grain, you'll see this same behavior in all breeds.
7. Block out times in your personal planner for routine maintenance such as hoof trimming, annual lice and tick checks, and body condition assessments.
8. Prevent illness related fatalities before they start. If you can spot a warning sign (such as low energy or lack of interest in food), you can take appropriate action before her health deteriorates further.

9. Know the common ailments of goats.
10. Whenever possible, keep a closed herd. Goat owners who are regularly bringing new goats in, participating in shows, and visiting other farms have a higher risk of stress-related and airborne illness in their own barns.

Annual Testing and Veterinarian Visits

If you live in a remote area, your vet may charge you for travel time on top of her standard rates. You can save a lot of money by coordinating vet visits with a neighbor or by having the veterinarian perform multiple tasks and tests per visit on your own farm.

What kind of testing? Standard blood tests for CAEV (caprine arthritic encephalitis virus), brucellosis, Johne's disease, and a skin or glandular test for tuberculosis.

Similar to vaccines, annual blood and skin tests may not be something you choose to participate in, especially if you have a closed herd. If you're increasing your herd, arrange annual tests and booster vaccinations to coincide with new doe arrivals.

You could save money and draw the blood for testing yourself and send it away by courier based on your veterinarian's instructions if the thought of poking needles into a live animal doesn't make you queasy. Have your veterinarian show you how to hold a goat and get a blood sample as well as how that sample should be packaged for shipment before attempting this task on your own.

While your veterinarian is present, send her back to the office with a fecal sample for laboratory analysis. The fecal sample is used to type internal parasites so that the most effective treatment can be prescribed. Ideally, you would have one sample for every goat, but why stress the budget? I send one sample to represent all goats that cohabitate. If you have two small groups of goats, or one main herd and one in quarantine, you would need two samples.

I just recently learned why these laboratory samples are so important. Over the years, we have all been taught to regularly worm our livestock with medication purchased from the feed store. This is half right. Medication can be purchased at the feed store, and doing so is an inexpensive task that helps keep our animals healthy. Where we've gone wrong is not knowing precisely what type of parasite we are treating without a laboratory sample.

We have also been taught to switch the brand of worming medication used every year. The concept behind this was that different medications target different types of parasites. By switching brands, we aim to get at least half of the parasites every year. Internal parasites grow resistant to type of worming medication after a few rounds of use—especially if we've been giving an incorrect dosage. Once these parasites are resistant to one drug, you can only hope to kill them off with a different medication, with a prayer that you are using the correct dosage this time around. Your laboratory sample and resulting medication ensures you are removing the full parasitic load at the correct dosage.

Aside from a standard parasitic load treated annually without harm to the goat, there are other types of parasites that could harm or kill your goat in under twenty-four hours. These parasites are commonly found in the digestive tract, but a few attack the liver and the lungs. Should you catch a warning sign, you may not have much time to react and correct the problem.

Preventative measures are worthy of your effort. There will always be parasites present on land and in your goat, but with proper management, you can ensure they don't get out of control. Second to incorrect dosage (which is based on weight—see "Weighing Goats"), stress due to improper care is the next highest cause of parasitic imbalance. The final two top contributors to infestation are ground feeding of hay (goats ingest expelled eggs in feces) and over grazing (goats being forced to eat short grasses on pasture).

Common Signs of Internal Parasitic Infestation

- Chronic coughing
- Diarrhea
- Dull coats
- Constipation
- Insufficient weight gain for feed quantity
- Lack of growth in kids
- Lethargy
- Partial paralysis
- Potbellies

Weighing Goats and Keeping Records

Weighing and recording your goat's weight every month is good barn practice. Weight is a benchmark to measure health, preparation for breeding, and kid development. Watching your goat's weight will also tell you if daily ration amounts require adjustments.

A sudden drop in weight is the earliest signal that health is waning, perhaps due to internal parasites if nothing else has changed in the goat's routine.

You can arrive at a quick, close approximation of weight using a measuring tape and the chart below. The measurement is taken from the goat's heart girth, which is the area directly behind their front legs. Goat weight tapes, which report the weight of your goat based on a correct heart girth measurement are available at the feed store. These two methods—the chart below and the weight tape—are susceptible to human error and should be used in conjunction with body condition assessment scoring.

To date, no reasonably close weight tape or chart is available to accurately measure the meat goat breeds.

If you are weighing kids with a heart girth of less than eighteen inches, you can use your bathroom scales. Weigh yourself first, then again with the kid in your arms. Subtract your weight from the second weight and you will have the kid's weight.

Weight measurement before medication is a vital necessity. Too much medication could cause an overdose; too little and the treatment won't be effective. To further complicate the matter, consistently underdosing literally trains organisms (bacterial and parasitic), to become resistant to the drug. With that resistance established, even a second correct dosage may not be effective.

DAIRY GOAT WEIGHT CHART

Weight Chart Large Breed Dairy Goats		Weight Chart Large Breed Dairy Goats	
INCHES	POUNDS	INCHES	POUNDS
18.25	23	34.25	120
19.25	27	35.25	130
20.25	31	36.25	140
21.25	35	37.25	150
22.25	39	38.25	160
23.25	43	39.25	179
24.25	51	40.25	180
25.25	57	41.25	190
26.25	63	42.25	200
27.25	69	43.25	210
28.25	75	44.25	220
29.25	81	45.25	230
30.25	87	46.25	240
31.25	93	47.25	250
32.25	101	48.25	260
33.25	110	49.25	270

Weight Chart Large Breed Dairy Goats	
INCHES	POUNDS
50.25	280
51.25	290
52.25	300
53.25	310

Weight Chart Large Breed Dairy Goats	
INCHES	POUNDS
54.25	320
For every 1/2 inch more . . .	. . . add 5 pounds

The Art of Drenching

Many medications require you to force-feed the substance to your goat with a process called drenching. When you drench a goat, you must ensure that the medication is inserted well into the back of the throat a little to the goat's left side, at the base of the tongue. Her head should be held slightly elevated but not so far that she cannot swallow. Never hold a goat's mouth open while drenching.

The first time you do this, it will go well, no trauma, no problems and you'll wonder why I've fussed so much over the instructions. Eventually though, usually on the second or third drench, your goat is going to take offence that you are shoving something down her throat and retaliate. From this point forward, you will need to be a little cunning to get the job done.

Here are a few tips on outsmarting a goat for drenching purposes:

- Bring a strong or stocky friend to hold the goat. I can't tell you how many times I've seen people topple over, get partly dragged, or simply pushed aside when their partner was about to insert a drench syringe.
- If the quantity is small, use a syringe without the needle to administer the medication. You can pry it in between the lips and slide it to the back of the mouth without cooperation from the goat.
- If quantities are larger or if you have many goats to drench, buy a drench gun or kit. These guns have a skinny nozzle on the end, making administration of medication a lot easier.

- Don't plan on being able to catch a goat that slips out of your grasp for at least an hour.
- Goats lurch forward during a drench. Be ready to move with her or you could damage the roof of her mouth.
- Drench right after a cud-chewing session. If you attempt to drench right when she's bringing up cud, you will be wearing both the cud and the medication.
- Don't put your face in front of hers. The regurgitated cud-medication mixture could end up in your mouth.

First Aid Kit

It took me a full year to wise up to the fact that keeping supplies in a goat barn was a good idea. After multiple trips back to the house I eventually assembled a goat first aid kit.

My kit is little more than a wild assortment of supplies crammed inside an old ammunition box, which I keep latched and tucked into the barn's rafters. Medications and vaccines are stored in our house refrigerator, and the Merck veterinary book sits on the shelf in my office. The book, I'll admit, is dusty, but having it gives me peace of mind.

Once you've raised goats for a few years, you'll have your own favorite supplies to keep on hand; but to get you started, here are the contents of my own barn kit:

- 3 clean towels wrapped individually in plastic bags
- antibiotic ointment
- antibacterial lotion
- baking soda
- electrolyte powder
- hoof shears and knife
- hydrogen peroxide
- hypodermic syringes and needles
- isopropyl (rubbing) alcohol
- mineral, olive, or peanut oil

- rectal thermometer
- rolls of surgical cotton, adhesive tape, and gauze pads
- udder balm

Electrolyte Replacement at Home

In a pinch for an electrolyte replacement for dehydration due to scours? Boil a quart of water first, then mix in 2 tablespoons of light corn syrup with ½ teaspoon of salt and ¼ teaspoon of baking soda. Allow to cool before using.

Health Records and Notes

Although you may not be one for keeping notes (and to be honest neither am I), forcing yourself to keep a barn journal may one day save your goat's life. A reference of a goat's change in eating habits, energy levels, or appearance is the best tool you can hand over to a veterinarian called in for assessment and treatment.

Barn records serve a secondary purpose. If you are ever called away or can't get back home to do chores, any friend or neighbor armed with these records could walk into your barn and take over with minimal risk of adverse effects. You'll find a health docket on www.KeepingFarmAnimals.com where you can record daily grain ration, vaccination schedules, changes in eating habits, quarterly weight and temperature, breeding dates, and more for every goat you own. Print as many as you like and keep them in the barn with a pen or pencil for reference or emergency.

By familiarizing yourself with the most common afflictions and illnesses, you'll know what to look for and can take immediate action.

You will get to know your goats so well that you be able to sense if one is even just a little off. These subtle changes, noted as they happen, can help you catch an illness, infection, or disease before it becomes life threatening.

I once had a doe with an eye that was tearing a lot in the middle of winter. I didn't worry about it because the fluid was clear and slight, and her mood and temperature were both fine. When the veterinarian came for another matter, I asked him about the eye, and the first question he asked was "How long has it been this way?"

Shocked, I couldn't remember. Had it been two weeks or two months since I first noticed it?

Thankfully, the tearing eye was nothing more than a mild sensitivity, but had it been worse, I might have carried guilt for a long time. This helped me to realize that a paper record is worth the few seconds that it takes, even if it never amounts to anything. More importantly, I learned that my memory could not be trusted when it came to the minor warning signs of waning health.

Vaccinations and Worming

Vaccinations are incredibly controversial and widely misunderstood. There are thousands of small farms across the United States and Canada who do not vaccinate their animals, but most of these farms keep closed herds for personal use only.

The first goats on my land were brought in for weed and brush control. I hadn't intended to use them for meat or dairy, to sell or breed, and they were never subjected to another farms' animals that might be carrying disease. My reasoning seemed sound at the time, but the more I learned about goats—and the more uses I intended them for—the more I began to realize how important vaccinations truly are.

Today I view vaccinations in the same light as I do insurance. A necessary expense that protects your investment and the lives of those you care about. Call a veterinarian, keep a dated record for each animal in your barn, and sleep easier at night knowing you've done everything in your power to keep your animals, and in turn your family, healthy.

Whether you perform the vaccination yourself or call in a veterinarian, you need to know about clear dates. A clear date is the recommended waiting period between vaccination and consumption of the milk and the meat of the animal.

Vaccinations aren't always a guarantee. Animals that have been vaccinated against a disease may not be fully immune to it. If your goat does contract or come into contact with the disease, her return to health will be quicker and the illness less severe, if previously vaccinated.

The top three to vaccinate against are enterotoxemia, tetanus, and chlamydiosis (the latter for breeding does only). You can give the vaccines yourself (usually available at most feed stores or with a working arrangement with your veterinarian) but have a veterinarian out for the first visit to least discuss the merits of each vaccine, and demonstrate both intramuscular and subcutaneous injections.

As a precaution when performing vaccinations yourself, pick up a small supply of epinephrine just in case your goat has an adverse reaction. In some areas, epinephrine is only available by veterinarian prescription.

Tips on Giving Vaccinations

- Only vaccinate a goat when she is in excellent health.
- Use a new, sterile, 20 gauge needle for each goat. (3/4 inch for adults and 1/2 inch for kids)
- Never use expired vaccines.
- Never mix vaccines.
- Read the label and follow the manufacturer's instructions.
- Write every vaccination down in individual health dockets.
- Dispose of used needles responsibly.

Here are the most commonly used vaccinations and the best time to give them for North America:

COMMONLY USED VACCINATIONS

Disease	Vaccine	Frequency / Time
Chlamydia abortion	Chlamydia	Bred Does: 4–6 weeks after successful breeding.
Cornybacterium pseudo-tuberculosis	CLA	Kids: 6 months old and 3 weeks later. All: Annual Booster
Enterotoxemia and Tetanus	CDT	Kids: 1–2 months old and again 1 month later. Does: 4–6 weeks before kidding. All: Annual Booster
Orf	Soremouth	All: Annually
Pasteurella multocida and Mannheimia Haemolytia pneumonia	Pneumonia	Kids and New Acquisitions: Two shots, 4 weeks apart. All: Annually.
Rabies	Rabies	All: Annually

Common Diseases and Illnesses

Some of the most common goat viruses and illnesses are listed below. For each listing, I've provided preventive measures when applicable.

In a few of the listings you'll find that there are no known cures. Please check with your veterinarian for recommendations of any animal's care. Thanks to modern science, treatments are discovered every year. The rising interest in goat ownership and herd management has served goat keepers well.

If you have any concern over the health of your animal, take a temperature reading, note any symptoms, isolate her in a quiet stall, and call the veterinarian.

ABSCESS

Firm to hard fibrous lumps under the skin of an animal caused by a bacterial infection of the lymph nodes. Most bacterial infections resulting in an abscess can be traced back to a flesh wound.

Prevention: Provide a safe yard, pasture, and shelter for your goats. Metal protrusions from the ground, fences, and barn walls can easily puncture a goat's hide. If you notice any wounds, treat them immediately and check daily for signs of infection.

Treatment: Isolate the goat from the herd and call the veterinarian to discuss severity. A surgical operation might be required to remove and drain the abscess.

ANAPLASMOSIS

An infectious disease with symptoms of anemia.

Prevention: Control of flies and ticks, presterilization of specialized equipment (namely hoof trimmers, castration devices, and the disbudding iron), and lowered stress.

Treatment: Intramuscular injection of tetracycline with veterinary examination and care.

ANEMIA

Lack of oxygen to body tissue. A deficiency of circulating red blood cells. The goat will appear to have a lack of energy. A whitened mucous membrane may show around the eye. Gums appear pale.

Prevention: Anything which causes a loss of blood—internal or external parasites or wounds. Sometimes brought on by blood diseases that are out of your ability to prevent.

ALLERGIES / INSECT STINGS

Severe reaction to new feed, insect bites or stings, or vaccines and antibiotics. Sudden collapse, coma, difficulty in breathing, excessive saliva flow from mouth or eyes, heart irregularities, mild to severe itching, hives, hair standing erect, or red areas on the skin.

Treatment: Remove the goat from the reaction location if you suspect the cause to be nearby. A treatment of an antihistamine may reduce the reaction. Consult with your veterinarian.

BLACKLEG

Caused by a specific bacteria living in the soil. Symptoms (fever, depression, loss of appetite) quickly lead to death.

Treatment: Antibiotics.

BLOAT

Bloat is a painful buildup of excess gas. If you see a growing swelling on the goat's left side and the goat is restless, don't hesitate to place a call the veterinarian's emergency number. Life-threatening if left untreated.

Prevention: A change in diet or too much ration is usually the culprit. Keep grain and ration well out of a goat's reach and introduce new foods slowly.

Treatment: While waiting for the veterinarian to arrive, you could try to keep the goat on its feet by employing another person, a wall of hay, or by any other means. Rub the goat's stomach to help move and release the buildup of gas. If no veterinarian is available, you could try to drench the goat (oral) with two cups of mineral, peanut, or olive oil followed with two ounces of baking soda dissolved in a cup of room temperature water.

BRUCELLOSIS

Abortion in last six weeks of pregnancy for does, swollen joints and testicles for bucks. Human beings are susceptible from contact with infected animals or by consuming milk products of an infected animal.

Prevention: Vaccination.

Detection: Part of annual veterinary blood test. Isolation and removal of any infected animal.

CAE / CAEV

This deadly virus—caprine arthritis encephalitis—is still without a cure. The virus is passed from one goat to another and is easily recognizable by swollen and stiff knee joints in mature goats and weak rear legs in kids.

Prevention: Annual blood tests. Only purchase goats that have a clear certification. Quarantine any new additions for at least a month and have your own vet test the goat before release.

Treatment: None.

CHLAMYDIOSIS (CHLAMYDIAL ABORTION)

This deadly disease is contagious to humans. If you are assisting with a live birth and the doe hasn't been vaccinated, wear long plastic gloves and scrub thoroughly after kidding. Warning signs are a premature abortion during the final eight weeks of pregnancy. If the doe carries the kids full term, they are either stillborn or very weak.

Prevention: Part of your vaccination schedule. Remove any affected does from the herd to prevent spreading the disease.

COCCIDIOSIS

A potential life-threatening parasitic infestation. You'll notice a lack of appetite and energy. Your goat may lose weight quickly or develop bloody droppings. Coccidiosis is caused by a protozoal parasite.

Prevention: Keep the goat house, feeders, and watering stations clean. Some feed may be available in your area with a coccidiosis deterrent, but it is not recommended for lactating goats.

Treatment: Coccidiostat available at most feed stores or through your veterinarian.

DIARRHEA / SCOURS

The trouble with scours and kids is that they very quickly become dehydrated and too weak to eat. Scours are often brought on by a change in diet but could be a sign of other problems.

Treatment: Isolate any kid with scours immediately. Disinfect every bit of your barn or goat shed. Call the veterinarian for advice and bottle feed kids with an electrolyte fluid (available at the feed store) instead of milk ration for two days.

ENTEROTOXEMIA

Classified as a life-threatening bacterial infection. This condition is also called overeating disease. Goats are seen to twitch, show signs of bloat, grind their teeth (a sign of pain in goats), and have an increase of temperature.

Prevention: Part of annual vaccinations. Avoid changes in diet.

Treatment: None.

FOOT AND MOUTH DISEASE (APHTHOUS FEVER)

Blisters on the mouth, tongue, teats, and in between the toes; will peel off and leave a pit in the goat's flesh. Lameness, abortion, kid death, and starvation are some of the symptoms.

Prevention: Vaccination for local strains.

Treatment: None.

HOOF ROT

If there was one good reason a goat's yard and bedding should always be dry, hoof rot would be it. This bacterial infection can result in death, but you'll notice and have time to treat it well before that happens. Symptoms are smelly feet, lameness, loss of weight. Tetanus is potential complication to hoof rot.

Prevention: Trim hooves regularly. Keep bedding clean and yard dry.

Treatment: Trim away hoof to a consistent length and to healthy tissue. Soak the foot for two minutes in water with with dissolved copper sulfate (gallon: ½ pound, respectively). If the condition is severe consult with your veterinarian for antibiotic treatment.

KETOSIS / PREGNANCY TOXEMIA

This metabolic disorder can be life threatening to a bred doe. Occurs most often a few weeks prior to, or just after, kidding. Feed quality that does not match nutritional needs causes the doe's metabolic response to draw energy from her own body to facilitate fetal growth. Warning signs are going off her feed, lameness, and/or sweet-smelling breath or urine.

Prevention: During the last half of pregnancy, there is substantially less space available within the doe's stomach for energy and nutrition. Her hay should be of the highest quality and offered as free-choice. Rations or any supplemental feeding should be specifically manufactured for pregnant does. Incorrectly fed bred does that were overweight before breeding are high risk, but ketosis is also common in first time mothers and does carrying multiples.

Treatment: Isolate the doe if you notice any of the warning signs. Entice her to eat by adding a feed-grade dry molasses mix to ration or grain or a treat of 2 parts corn syrup with 1 part molasses (20–30 ounces every two hours). Monitor her fluid intake. With enough food energy and sufficient water, she may be able to flush the ketones out of her kidneys naturally. Ketones are the byproduct of a doe's metabolic system converting body fat into glucose for energy. If you can't entice her to eat, call the veterinarian immediately.

LAMINITIS (FOUNDER)

A secondary complication of another illness, improper hoof care, or a change in feed. The goat's hooves become inflamed, and she may walk on her front knees. The coronary band area becomes inflamed and fevers are often noted.

Prevention: Aside from the obvious (proper hoof trimming, progressive feed changes, and good health), a clean living arrangement is required to prevent Laminitis.

Treatment: Antihistamines and electrolyte drink followed by antibiotics.

LICE

Consistent scratching and biting, loss of weight, loss of hair, and decreased milk production could be warning signs of a lice infestation. Lice can be an ongoing problem when goats are living in damp climates or shelters.

Prevention: Keep living quarters dry. Avoid contact with lice-infested animals.

Treatment: Any powder, dip, spray or pour-on insecticide approved for livestock or dairy animals will work and should all be available at your local feed store.

MASTITIS

A painful bacterial infection in a doe's udder. The udder may be hard, swollen, or abnormally hot or cold. Milk might smell bad or show signs of blood. Infected does may stop eating.

Prevention: Keep living quarters clean, dry, and safe. Apply a teat dip after every milking session and try to keep her standing for thirty minutes after milking. Once a month, check your does with a mastitis home test available from your local feed store. Dairy cattle tests can be used with accuracy for goats.

Treatment: Isolate the infected doe. Milk three times a day to relieve pressure. Apply hot packs to the udder four times a day. Dispose of all milk. If your veterinarian prescribes antibiotics, be sure to record the clear date.

MILK FEVER (PARTURIENT PARESIS)

A blood calcium deficiency often noted in does around kidding dates. She may drag her back feet or be incapable of standing for labor and delivery. Constipation is also noted. Complications at time of kidding may include enterotoxemia, mastitis, and a retained placenta.

Prevention: Maintain a suitable intake of calcium without disrupting the natural rhythm of a doe's health. Don't overfeed alfalfa

Treatment: Veterinarian assisted intravenous calcium supplement during heart monitoring.

MYCOPLASMA

Mycoplasmas are actually microbial organisms that may exist in an animal without any indication or complication. They are neither bacterial nor viral. Kids under stress are most likely to suffer from an infection spread by mycoplasma present in a dam's colostrum. Symptoms are often mistaken and treated as pneumonia, mastitis, polyarthritis, pinkeye, or spinal cord disease, which can also be complications of mycoplasma.

Prevention: Pasteurize all colostrum when bottle feeding (133–138 degrees F for one hour without over heating). If a doe is suspected of having mycoplasma due to previous kid deaths, remove future kids from her and bottle-feed.

Treatment: No vaccine.

NAVEL (OR JOINT) ILL

Contamination of the umbilical cord after birth. Could cause lameness, internal abscess, and physical retardation.

Prevention: Keep kidding stalls clean. Perform proper sterilization (7 percent iodine dip) of the cord after cutting.

ORF

Also known as sore mouth, scabby mouth, contagious pustular dermatitis, sheep pox, and contagious ecthyma. This viral condition causes thick scabby sores on the lips and mouth, which could last from one to four weeks. Lesions can also appear in the nostrils, around eyes, on thighs, vulva, and udders.

Quickly manifests as an outbreak among all animals copasturing or cohabitating. Kids may have trouble nursing and suffer malnutrition. Nursing kids can pass the infection onto the teats of a doe, further escalating into mastitis.

There is a human health risk associated with both the virus and the vaccine.

Prevention: There is a live virus vaccine available, but it is seldom suggested for use by farm veterinarians. The best prevention is to keep a closed herd, wash thoroughly after visiting other farms or animals, and don't show your goats.

Treatment: Very little can be done as this is a viral infection that must run its course. Consult with veterinarian to treat deviated manifestations (sores on eyes, teats, bacterial infected pustules, etc.). Follow a strict sterilization schedule for at least four weeks after the first sign of orf. An after-infection, veterinarian-supervised, vaccine is now available to minimize the severity of further incidences.

PINKEYE

Can be caused by both bacteria and virus. Goats will squint and have watery eyes.

Prevention: Contagious. Avoid contact with infected goats. Can also be transmitted by flying insects and dusty conditions.

Treatment: The entire herd should be treated with antibiotic drops, ointment or powder.

PNEUMONIA

Various bacteria and viruses attack an already stressed or ill goat. May also be brought on by unrelated allergic reactions. Warning signs are coughing, loss of appetite, fever, and runny nose and eyes.

Prevention: Keep your goats as stress free as possible. Ensure their housing is dry and free of cold drafts.

Treatment: Call your veterinarian for antibiotic treatment.

RABIES (HYDROPHOBIA)

A bite or saliva from an infected animal spreads the rabies virus. In goats, symptoms can take anywhere from ten days to seven months to appear. Untreated symptoms can result in death within two to twelve days. Symptoms range from frothing at the mouth, aggressiveness or other change in behaviour, a change in voice, paralysis, and no interest in food or water.

Prevention: Annual vaccination. Keep wild or stray animals away from the herd.

Treatment: If you suspect a rabid interaction, isolate your animal and sterilize any wounds. Your veterinarian will then begin a series of vaccinations to protect the goat before symptoms appear.

RESPIRATORY TRACT DISEASE

Germs, which are naturally occurring in the respiratory tract of goats, attack and cause illness when an animal's immune defense is weakened.

Prevention: Stressors that contribute most to respiratory tract disease include lack of ventilation or wet living conditions, drastic changes in feed or climate, long distance hauling, overcrowding, and parasitic infestation.

Treatment: Veterinary consultation.

RINGWORM

Ringworm is brought on by fungi in the soil that attacks the skin of an animal. It shows up as circular skin discolorations and hairless patches on the head, nose, neck, or udder.

Prevention: Highly contagious to both animals and humans. Avoid contact with infected animals.

Treatment: Wear rubber gloves and follow a strict sterilization practice. Scrub the patches with warm soapy water and coat with iodine or fungicide available from your local feed store.

SALMONELLOSIS DYSENTERY

Often triggered by high stress conditions. Symptoms include black or bloody scours, unexplainable kid death, or premature abortion.

Prevention: Stress-free and clean living conditions.

Treatment: Veterinary consultation recommended.

SCRAPIE

Scrapie is a neurological disease (similar in nature to Mad Cow disease) that has been studied since 1952 with no known cause or cure.

Only seven cases have been reported in goats and the disease in all cases had been acquired from contact with sheep. Further information on scrapie on www.keepingfarmanimals.com, or (if in the USA) by calling the USDA Animal and Plant Health Inspection Service.

Prevention: None known.

Treatment: Call the USDA at 866-873-2824.

TETANUS

Harmful bacteria enters through a flesh wound and creates stiff and spasmodic muscles. Early warning signs are wide eyes and flared nostrils, escalating to muscle spasms and eventually death.

Prevention: Part of a kids' vaccination schedule at four weeks and then eight weeks of age.

Treatment: Veterinary care may be practical for valuable animals.

TICKS

Goats foraging in wooded areas may have a problem with ticks. Carrying and spreading a wide variety diseases, ticks are also an annoyance to your animals. You may discover your goat rubbing or scratching the tick's point of entry. Other symptoms include hair and weight loss.

Prevention: Check forest-pastured goats daily during tick season and remove them immediately.

Treatment: If discovered early enough, a tick can be removed by a gentle rubbing in a circular motion with your finger. The tick should let go in under a minute; if not, rub in a counterclockwise direction. Many people tug on tick's bodies with a pair of tweezers; this sometimes works, but as the tick is a burrowing insect, you run the risk of leaving the head (and potential disease it could be carrying) inside your goat's skin. Talk to your veterinarian.

TUMORS / WARTS

Seen more often in the Angora and Saanen breeds, but any goat can have tumors, warts, or warts that grow into malignant tumors.

Treatment: Surgical removal by veterinarian if necessary.

URINARY CALCULI

Formation of mineral stones in the urethra of a goat. Most troublesome for bucks and wethers as the stones cause blockage in the urinary tract.

Prevention: Avoid early castration. Balanced diet and minerals. Calcium to phosphorus ratio should be maintained at a ratio of 2:1.

Treatment: Veterinary consultation, often resulting in expensive surgery.

WOUNDS

Goats are prone to cuts and scrapes due to their curious and adventurous nature. Although the cuts themselves aren't much of a problem the introduction of diseases and infections through flesh wounds can be life threatening.

Treatment: Stop any bleeding by applying pressure with a clean towel. Sanitize the wound area immediately with hydrogen peroxide. Clip hair around the wound and flush the area with warm soapy water, then rinse with clean water. Apply hydrogen peroxide again and cover the area with an antibiotic ointment. Check, clean, and redress the wound daily until it has fully healed. If the wound becomes infected, call your veterinarian.

Poisonous Plant Ingestion

There are many plants, weeds, vines, and brush that are poisonous to goats. While some are common to specific regions and in particular seasons, the chances of a well-fed goat ingesting these plants are minimal.

The average goat will not eat a poisonous plant if there is ample grazing, forage, and hay available to them. They will graze around the plant and may partake of it, if and when it has grown past its toxic stage. This, however, does not take the onus of responsibility off of us to do our absolute best to ensure the goat does not have access to these plants.

Many plants—vegetable and floral—that we showcase in our home gardens are not familiar to a goat's instinctual survival, just one more point in favor of adequate fencing. Add to this that goats are naturally curious and explore the world orally and you can see why it is so important to ensure they are well fed and well contained.

A goat that has partaken in a poisonous plant may show one or more of these signs:

- blue mucous membranes
- cold extremities
- coma
- depression or nervousness
- diarrhea or constipation
- head thrown back and rigid
- heavy breathing or respiratory distress
- loss of coordination, staggering, paralysis
- loss of rumen function
- muscle tremors or convulsions
- red oral tissues
- salivation or frothing
- teeth grinding
- vomit
- weak pulse

Potentially Poisonous Plants

Plants known to be toxic or poisonous to goats in varying degrees of severity are:

- Aconite
- African Rue
- Allspice
- Andromeda
- Arrow Grass
- Avocado
- Azalea
- Bagpod
- Baneberry
- Belladonna
- Black Snake Root
- Bloodroot
- Blue Cohosh
- Boxwood
- Broomcarn
- Buckeye
- Buckwheat
- Burning Bush Berry
- Buttercups
- Calotropis
- Cassava
- Castor Bean

- Celandine
- Cherry
- China Berry Tree
- Choke Cherry
- Clover
- Cocklebur
- Coffee Weed
- Common Poppy
- Corn Cockle
- Crotalaria
- Crow Poison
- Crowfoot
- Datura
- Death Camas
- Dicentra
- Dogbane
- Dog Hobble
- Downy Broome Grass
- Dumb Cane
- Elderberry
- Euonymus Bush
- False Hellebore
- False Jessamine
- False Tansy
- Flixweed
- Fume Wort
- Fuschia
- Goat Weed
- Ground Ivy
- Hellebore
- Hemp

- Holly Tree / Bush
- Horse Nettle
- Ilysanthes Floribunda
- Indian Hemp
- Indian Poke
- Inkberry
- Japanese Pieris
- Japanese Yew
- Jimson Weed
- Johnson Grass
- Kafir
- Klamath Weed
- Lantana
- Larkspur
- Lasiandra
- Laurel
- Leucothoe
- Lilac
- Lily of the Valley
- Lobelia
- Locoweed
- Lupine
- Madreselva
- Maleberry
- Marjiuana
- Maya-Maya
- Milkweed
- Milo
- Monkshood
- Mountain Laurel
- Oleander

- Pieris Japonica
- Pine Trees
- Pink Death Camas
- Poke Weed
- Poison Darnel
- Poison Hemlock
- Poison Rye Grass
- Ponderosa Pine Needles
- Purple Sesban
- Ranunculus
- Rape
- Rattlebox
- Rattleweed
- Red Maple
- Rhododendron
- Rock Poppy
- Sand Bur
- Senecio
- Sevenbark
- Snakeberry
- Sneezewood
- Soapwort
- Sorghum
- Spider Lily
- Spotted Cowbane
- Spotted Water Hemlock
- Spurge
- Squirrel Tail Grass

- Stagger Grass, Weed or Brush
- Sudan Grass
- Sweet Shrub
- Thorn Apple
- Varebells
- Velvet Grass
- Water Hemlock
- White Cohosh
- White Snakeroot
- Wild Black Cherry
- Wild Hydrangea
- Wild Parsnip
- Wolfsbane
- Yellow Jasmine
- Yew

Should you suspect that your goat has been poisoned, the veterinarian isn't answering her phone, and you are miles away from a feed store, there are a few home remedies that might save your goat's life.

1. If you have caught your goat eating a large amount of a toxic plant you could induce vomiting by pouring 2 tablespoons of table salt into her mouth and holding her mouth closed until you are certain she has swallowed the salt. Within a few minutes she should vomit.

2. Activated charcoal absorbs harmful toxins in a goat's gut. Follow quantity and instructions on the box based on goat's weight.

3. Here's an old-timer's recipe to stimulate rumen activity and flush out toxic plant matter. Your goat won't drink it on her own, you'll need to drench the fluids into her.

2 cups warm water
1 tsp ground ginger
1/2 tsp baking soda
1/2 tsp salt
1 tsp molasses
1 tbn epsom salts

Yield: serving for 1 adult goat. Adjust quantities for younger goats.

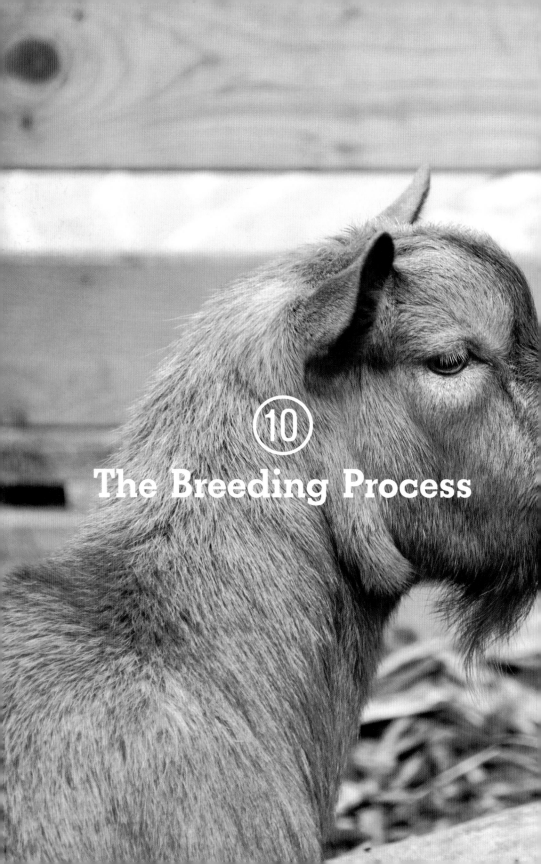

10
The Breeding Process

THE GOAT IS a seasonal breeder.

As the days grow shorter, the doe begins to look for a suitable mate. Scientifically, we say she is seasonally polyestrous, but in layman's terms, we simply call her a short-day breeder. As a rule, she won't be very picky about breeding either. This makes her owner's task of finding a suitable mate easy to accommodate.

Aside from choosing a buck for her, there will be equally important decisions to be made—namely when she's bred, the best time to have kids, and what to do with all those extra mouths to feed.

Animal husbandry is never an exact science; the act of procreation even less predictable. As I write this chapter, I can think of a gross exception for every time span I've quoted, every breed I've mentioned, and every situation I've warned against. What follows is to be considered a guideline that loosely covers the many breeds and classes of North American goats.

Unless you plan on only growing wethers that you purchase from a local farm, you need to be well armed with knowledge, and in control of, the breeding process. From buck selection criteria to due dates, being in control of your doe's breeding is as important as her supply of daily fresh water.

The typical breeding season runs from September to February. At our farm in central Ontario, I hold off breeding does until late October to early November with an expectation of kids in late March and early April.

Late and staggered breeding dates result in late and staggered due dates across my small groupings. There have been times when a few kids are born in the same evening, but I usually have time to carefully watch over each doe while she kids. Where spring takes its time arriving, breeding late is precautionary. Kids are not born into a subzero Fahrenheit barn months before the pasture is ready for them. Late breeding also saves me money. By the time kids are ready to eat solid food, it will be growing just outside their front door.

ABOVE Although goats are only slightly uncomfortable in temperatures just below freezing, it will be easier on your farm budget if kids reach their wean date after the snow has left.

When to Breed

Even though your doe might think she is ready to breed at six months of age, don't allow it until she is eight to nine months old. A better rule of thumb is to wait until she has reached at least 75–80 percent of the average mature size for her breed. Early breeding could result in a stunted doe that has trouble kidding. In subsequent years, she will also have less chance of multiple births.

The other option—letting her miss her first year of breeding—fares less well according to some breeders. Waiting a full extra year will not only cost you money without return, but it will also create a doe that is a poor milker and a nonattentive mother. They advise that even if she hasn't hit the right weight on the scale, breeding young is the better option.

Dairy goats are bred every year. In the warmer regions of North America, does are bred during September and October. Then after a five-month-long gestation, they give birth right before spring, when fresh pastures will soon provide the most nutrition.

Meat goats are usually bred as often as possible, every eight to nine months, with the optimum breeding time between August and October.

Angora goats are bred, like dairy does, once a year. This is usually between August and November, right after a fall shearing. By arranging and controlling this goat's breeding, we are assured that they'll be ready to kid shortly after the spring shearing.

Goats raised for cashmere are bred anytime before November. This practice helps all owners to wean kids by mid-June and dry the does off. Lactation, it seems, slows the annual growth of new cashmere.

Controlling a Doe's Cycle

When you control the estrous cycle as well as the breeding of a doe, you become the master of your herd. Not only will you be controlling the approximate due date for kids, you'll also control the time that milk flows once again from your dairy does.

To manage this, all does should be kept securely, and a good distance apart, from bucks. You simply can't run a buck with your does and maintain control of breeding. In many cases you can't even keep a buck within two hundred feet of does and maintain control. The mere scent from the glands behind a buck's horns can induce estrous.

Knowing this, you are now also in control of when a doe will come into heat (the estrous cycle). More or less. We can't deviate too far from instinct after all.

To induce estrous in does, we could allow the sight and smell of him through a secure fence or employ the use of a buck rag. Once a buck has been noticed, most does will follow with a heat in seven to ten days.

Why Control Breeding and Estrous Cycles?

1. Does bred too young are at high risk for difficult births and have decreased ability at future multiple births.
2. If you control breeding and estrous cycles, you also control milk flow and quantity.
3. Kids born after the frost has left the ground will be on pasture sooner, expend less energy on maintaining body heat, and will grow quicker.

Inducing Estrous

If a month has passed into the breeding season and your doe hasn't exhibited any signs of interest, don't worry. Signs of the first (as well as the last) heat of the season can be subtle. Some does may never show a sign during these two heats. These are called silent heats.

If you don't own a buck and can't borrow one before a planned breeding, but you do want to speed up a doe's cycle, try the old farmer's trick of exposing your doe to a buck rag—with, however, my modern twist.

It has been a long held belief that the scent of the buck will induce estrous, but today's science tells a different story. It is not the scent, but the pheromones (closely related to scent) that cause the reaction.

ABOVE Does become more vocal when they're in heat, calling out for hours on end to raise any nearby buck's attention.

With mindful collection of scent, you may be able to pick up the essence of pheromones, by rubbing an old kitchen towel or facecloth on a buck. Spend most of your effort at the base of his horns and work the rag along his entire back. When you return home, allow your doe to smell and investigate the rag twice a day. Store the rag in an airtight jar when not in use.

ABOVE Choose your buck based on genetics and past performance. This spotted Boer from Rebel Ridge Farms in Virginia was fashioned through an intelligent, managed breeding program.

Is Your Doe in Heat?

Your doe may not exhibit all of these behaviors, and she may even have a few of her own.

- Vocalization. Some will bleat loudly as if in pain.
- Wagging the tail more than usual.
- The under tail area may darken, appear swollen, or be wet with mucus.
- Dairy does will have a change in milk volume. High volume right before the heat and less volume during the actual heat.
- Could show a decrease in appetite.
- May show an increase in urination.
- Will pace fence lines, sometimes to the point of exhaustion.
- May mount other does or let other does mount her.
- May have a slightly swollen and reddened vulva.

Knowing When It's Time to Breed

Until successfully bred a doe will continue to come into estrus every eighteen to twenty-two days. Each estrus will last from twenty-four to forty-eight hours. As soon as you notice her in heat, you will only have a small window of time in which to breed. During her heat she will go through a standing heat when she will stand still and calm as a buck mounts her. Standing heat can last anywhere between six and twenty-four hours.

If you have been away from home or think you have missed the standing heat, your doe's vaginal discharge may give you a clue of where she is in her cycle. At the beginning of the one- to two-day heat, her discharge will be clear. This progressively changes to a whiter shade of opaque by the end of standing heat.

Ovulation occurs twelve to thirty-six hours from the start of a standing heat. The average life span of buck sperm in a doe's genital tract is twenty-four hours.

The Best Buck

Finding a suitable buck could be the largest challenge in your managed breeding program.

If you are only breeding to freshen a dairy doe and aren't planning on keeping the kids or selling purebred registered kids, any healthy buck of any breed will do. As far as size is concerned, if you aren't concerned about raising large kids for freezer meat, pregnancy and kidding will be a lot easier on your doe if you can find a small buck to breed with.

Breeding for Genetics

- A dairy breed buck will pass on the genetics of his dam if a doeling is produced.
- A polled dairy buck, bred with a polled dairy doe, creates offspring with a 50 percent chance of infertility.

When the breeding program revolves around registered does, or when the objective is to cross breed for specific traits, you'll have a true quest on your hands to find that perfect buck. The same holds true if you plan on increasing your herd with the offspring as you'll want the best buck you can afford to employ.

Artificial Insemination

Artificial insemination (AI) is relatively new to goat breeders but saves the hassle of keeping a buck or transporting, and possibly stressing, animals. You'll be able to pick and choose amongst hundreds of available registrations to find the exact traits required to improve your herd from one breeding to the next. At the moment Google has 91,000 listings. Since the insemination kit can be shipped by mail, you won't ever be short on options.

Although I have heard of some small farmers performing the insemination themselves, most will find and employ an expert for such a delicate matter. If you think AI is your best option, find a practitioner in your area who will source out a service catalog of potential matches. Veterinarians are also a great resource, as are goat husbandry magazines.

The AI practitioner or your veterinarian can assist you in choosing a buck that has all the qualifications you seek to compliment your doe, and he will obtain and store the semen at his facilities. When the right time comes, he will inseminate your doe. Some of these practitioners come directly to your farm while others will require that you transport the doe to their office.

Sizing Up Potentials

If you are fortunate enough to have a network of goat farmers in your area, you may have a few suitable bucks to choose from. Bucks are selected for more than their breed, adult size, or coloration. Buck genetics pass on many positive traits to offspring such as milk output of his dam, potential for multiple births, and climate hardiness.

Find out as much as you can about any buck you are considering for breeding, including his past breeding performance and ancestry.

While you're at the buck's farm, have a look at his housing, do your best to judge his disposition, and definitely assess his current state of health.

A buck with a dull coat, very thin hair or with patches of hair missing may have lice, ticks or ringworm. Any or all of these will be passed onto your doe upon breeding.

Steer clear of bucks with poor confirmation, bad legs or improper hoof care—no matter what registration or paperwork is shown and even if you only plan on raising the kids for meat. If he doesn't look healthy and vibrant to you at breeding, neither will his offspring.

A buck should be in excellent health as well as worm and parasite free. He should have been on a concentrate ration for at least two months prior to breeding. Ask to see the buck's immunization and vaccination schedule and any veterinarian records for the past two years. Never take a stranger's word that his buck has a CAE/CAEV clear certification. You need proof or the life of your goats at home could be affected forever.

Best Prebreeding Practices for Does

Worm a doe one month before breeding. Every ounce of nutrition she ingests after breeding and gestation should serve her and the developing kids she's carrying.

If she is a standard weight or slighter with little to no fat, increase her feed quantity slightly. The term for this is flushing. Flushing a doe before the breeding season moves more eggs from her ovaries ready to be fertilized. She'll have a better chance of becoming pregnant and is more likely to have multiple births. Flushing will not work on overweight does. In fact it will hinder their chance of a successful labor and delivery.

One Month Before Breeding

Dairy breeds are flushed by increasing their daily grain ration. Meat breeds can be flushed by moving them to fresh and nutritious pasture for a month prior to breeding. If you don't have access to fresh pasture, you can increase the quantity of, or introduce, a corn ration. Keep meat does on lush pasture or with extra corn and dairy does on the extra grain for four weeks past breeding and then slowly decrease the quantity to their standard ration amount.

The Breeding

There are many ways to breed a buck and doe, but usually once you've provided the buck and the venue, your work will be done. Watch over the two and record the dates of successful breeding. If you can give does and bucks a week or more together you'll have a better chance of success.

Leaving a buck with a small herd of does for two complete estrous cycles (thirty-six to forty-four days) is common as long as the buck doesn't have another breeding to attend to. One buck can effortlessly service and impregnate twenty does over two cycles.

Also common are what I call driveway breedings. One goat is brought to another, and with the buck harnessed and leashed, he mounts and breeds the doe. The act is often repeated later in the same day. The success rate of this style of breeding relies heavily on the doe's owner, knowing precisely when she is in a standing heat.

A doe who is not ready to be bred will repeatedly move away at every buck's advance. I have also heard of a doe not liking a buck well enough to let him breed her. Some will be shy and won't breed with an audience. Run shy does with a buck for a few weeks to be sure she's bred.

Shy, didn't like the buck, or simply not ready—all excuses aside, a buck should mount a doe several times during contact. The more often you notice and take note of each breeding in your barn records, the closer you'll be in an estimated kidding date and understanding your doe's cycles.

If you have borrowed a buck and you will be registering the kids, ask the buck owner for a service memo. The information you'll need for registration after the kids are born are the date of breeding, the names and registration numbers of both buck and doe, both owner's names, and the buck owner's signature.

The Expectant Mother

A successfully mated doe is said to have settled. She'll sleep more. She'll eat more. And after 145–155 days of gestation, she'll gift you with a kid or two. Maybe more.

Some does act like they're in heat again even though they've been successfully bred. Frustrating to experience while you wonder if the breeding missed and if you'll have to wait another year for kids or milk.

Watch her carefully (Is she eating more? Is she sleeping more?) or have a vet come in to check her by ultrasound six weeks after breeding. There are also blood tests that can be performed and a milk strip test to tell you if she's faking the heat but is actually impregnated.

Eight Weeks Before Kidding Date

Dairy does that are still producing milk are usually dried off two months prior to their due date. Drying off a doe is a natural, uncomplicated process. The doe, no longer being milked twice daily, will stop producing milk, and the food she eats will now be converted into energy that serves the kids she's carrying. She may even put on weight.

When you dry her off, you must do it instantly—do not dry her off gradually, thinking it will be easier on her. Does dried off gradually have been noted to develop fibrosis, lower milk production in later cycles, and a higher risk of mastitis. After a week of not milking her, if she looks terribly uncomfortable, you can take *some* of the udder pressure off. Only do this once. Although opinions vary, I don't suggest milking her completely as it may encourage further production of milk. Just take enough off to relieve the pressure.

During the two weeks, it will take her to return to a comfortable state of a dry doe, gradually decrease her daily grain as it also slows milk production, therefore easing udder pressure.

Four to Six Weeks Before Kidding Date

Four to six weeks previous to kidding talk to your veterinarian about a vitamin and mineral booster of A, D, E, and selenium.

Vaccinate against overeating disease (enterotoxemia) and tetanus four to six weeks before kidding. Some of the doe's immunity will be passed on to her newborns. Some veterinarians believe improves the protective value of colostrum (first milk) against the two diseases for newborn kids.

One Week Before Kidding Date

Remove the doe from the herd, trim her backside and udder of any hair, and move her into a quiet stall where she can kid in peace and where you can keep a watchful eye over her.

New mothers and does having multiple births often kid early. Keep a close eye on does that are late to kid. They may have extra large kids coming and may require assistance.

As the time draws near, you'll be able to see and feel kids moving on her right side. Twelve hours before they're born, they'll stop moving—you can almost set your watch to the delivery hour at that point.

Other signs that kidding is imminent:

- As the kids move into the birth canal, the roundness of your doe's belly shifts. If you spend a lot of time with her, you'll notice it right away.
- If she's having multiple births, her tail may go up and stay up.
- She will have a mucus discharge on her back side.
- She may paw at the ground. She may repeatedly lie down and stand up. She will appear to not be capable of being comfortable.
- She may display extra affection to you by rubbing against you and talking to you.

Assisting with the Birth

There is no greater joy than to watch the miracle of life taking place. Waiting for the newborn to emerge is charged with anticipation, but it is the first breath of a new life that always brings me to tears. Whether the newborn is a human child, kitten, or a goat I'll gladly forgo a night out on the town, dinner at my favorite restaurant, or both, in hope of catching that very moment.

Being present for the birth also ensures that you can step in, if needed. Seldom do problems occur, but you'll want to be prepared with a kidding supply box, just in case.

I keep my kidding supplies in a large laundry hamper by the back door, ready to get to the barn an hour or two before I think kidding will take place. Inside you'll find:

- clean towels,
- a stack of old newspapers,
- a large cardboard box,
- iodine and a small cup,
- a bathroom scale,
- a few blank notebooks (one for each kid born to record delivery notes, weight and time),
- a newborn sized baby bottle and nipple, and
- a blow dryer and a heat lamp if nights are still chilly.

If you thought of it the last time you were at the feed store, you might have a few bottles and some lamb nipples in your basket, but if you forgot, don't fret. A human baby bottle with an enlarged hole in the nipple works equally well. (If you don't have one, I bet you can think of a neighbor who has a baby or grandchild who would be happy to help you out.)

The kidding process usually follows in this manner:

- Doe begins to strain in a laying position.

- A water sac shows then breaks. Two hooves appear and then a nose. A few more pushes and the kid is born.
- The doe licks her kid, and the first breath of air will be taken into the kid's lungs.
- A second, third, or even fourth kid may follow.

Most kid births happen without interference from humans, but there are ways to help without being invasive. If you have spent lots of time with your doe in the past, she will appreciate you being there, even if all you are doing is softly talking her through the steps. My daughter Veronica will sing to our animals, and I've seen the calming change this brings in many stressful situations.

Here are some ways to assist in the delivery process:

- Cover the wet birthing area with fresh bedding or newspapers to absorb the fluids and keep the stall dry.
- If the second kid is coming before the doe has fully cared for the first, dry off the kid with clean towels and gently wipe the kid's nose and mouth. If the kid has not yet taken its first breath, gently tickle the inside of his nostril with a soft piece of hay.
- A kid's first breath should sound clear—not rattled or bubbly. If you sense that fluids are in the air passages, lift the kid's back legs a few inches off the floor for ten seconds at a time.
- If your doe is distracted with a consequent birth and if the weather is unseasonably cold, you can use a hair dryer on a low, warm setting to dry a kid off. Once dry, place the kid into your cardboard box placed near a heat lamp (not directly under) to ensure extra warmth is available. Once your doe has finished kidding, place her kid at her side and remove the box from the stall. I also suggest removing any heat lamp from a goat's stall, but if you've allowed your doe to breed too early in the season and there is a chance that the kids may freeze, that can only be your decision.
- Tie off the umbilical cord with a soft string three to six inches from the kid's body. Sever it at the doe's side and spray or dip the end with iodine.

You can help your doe regain her strength after the kids have been born and are settled by providing a few tablespoons of cider vinegar mixed into warm water. Livestock molasses is another mix option. Ensure she has free-choice hay and fresh water.

As an extra precaution, once dam and kid have bonded, dip the hooves of newborn kids into iodine. Soft hooves absorb bacteria almost as easily as umbilical cords.

Twelve hours later, your doe will pass afterbirth. She may attempt to eat it. This is instinctual behavior (to save predators from finding her and her kids), but unnecessary in stall births. You can wrap it up in newspapers and dispose of it. If afterbirth does not fully emerge, call the vet—do not attempt to assist with this, or you'll risk losing your doe.

Birthing Positions

I will never let a doe struggle in labor for over thirty minutes without a sign of progress. If this is your first live birth, friends and neighbors who have livestock make helpful allies. Your veterinarian should also be on call for advice and an emergency farm visit no matter what hour of the day it is.

Colostrum

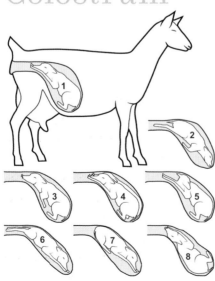

LEFT There are seven birth positions during labor. The two most common—front legs and head first, rear legs first— will require no assistance on your part. Other positions may require your help to move the kid—still in the birth canal—into proper position. Have a goat friend or veterinarian on speed dial and keep long rubber gloves plus lubrication in the barn when kidding is imminent.

The first milk a doe lets down after birth is called colostrum. Colostrum is loaded with extra nutrients, fats, and antibodies that newborns need.

If your doe's underside and udder is messy after kidding, clean and dry her udder before kids start nursing. The value of colostrum decreases exponentially within six hours of kidding. For optimum immunity, kids should ingest ten percent of their birth weight in colostrum within the first twenty-four hours of their life.

Storing Colostrum for Later Emergencies

Once all the kids have eaten and are cared for, you can milk your doe of remaining colostrum. This can be frozen for later emergency use. You may never need it, but it could save a small life if you lose a doe during kidding or need to feed a newborn kid by bottle.

Collect and freeze the colostrum as early as possible after kidding. Freeze in new ice cube trays, remove from the trays as soon as frozen, and store in an airtight container or vacuum sealed bag. Thaw colostrum at room temperature or at a low temperature without overheating. Colostrum can be pasteurized at 131 degrees F (56 degees C) for one hour, but unless the doe it was collected from is a known carrier of CAEV, most won't bother to pasteurize.

The Nursing Kid

As does are generally dried off for two months prior to kidding, her teats may be blocked with a waxlike plug. Nursing kids will dislodge the plug with persistence, but you can milk a quick strip yourself to make life easier on the kid and dam.

Within two hours, every kid should feed. If the kid has not latched on by this time, you can hold him near the teat and squirt a small amount into his mouth. You could also try bottle feeding him after cleaning the doe's teats and extracting the milk into a baby bottle. Call the veterinarian if you need a professional opinion. Kids who will not latch on or be fed by bottle will need to be tube-fed.

In dairy goats, it is common to immediately remove kids from their mother's side. This is done to protect dairy teats from aggressive feeders and to facilitate the practice of weaning kids at a younger than natural age.

Bottle-feeding is said to result in tamer, more human-friendly, adult goats. None of my dairy does started out life as a bottle baby, and I can't imagine any of them tamer than they already are. This decision is highly personal and will be based on your farm strategy—do you want to keep a freshened doe's milk for household use or sale, or would you be content sharing with the kids after a month, knowing that a natural start is the healthiest option for them?

Bottle Feeding Basics

Bottle feeding directly from birth requires your commitment to the barn every two to six hours per day, every day, for about three months. In the past, if you had many kids and a few helpers, there were products and methods to cut down on the hours spent, but not the frequency of visits. That was then. With modern technology, a "super nanny" automated feeder is capable of serving perfect-temperature milk for up to thirty-five kids at a time. When you factor in the potential years of use and the time it will save you the cost is more than reasonable.

Newborn kids require feeding every few hours with milk warmed to 100 degrees Fahrenheit. By the time they are two to four weeks of age, they require feeding every four to six hours. Kids will consistently consume 15–25 percent of their own body weight daily until they are four weeks of age.

KID BOTTLE FEEDING SCHEDULE

Age	Amount of Milk (oz)	Times / Day
1 to 3 days	4	8 to 12
4 to 14 days	8 to 12	6 to 8
2 weeks to 3 months	16	4 to 6
3 to 4 months	16 ounces	2

Start the kid on her dam's milk whenever possible. If you can't collect milk from the dam and you have colostrum on hand in the freezer from an earlier birth, use it for the first three days of feeding. Calf or lamb milk replacer may be fed from the fourth day. Check the content label and ensure it is at least 20 percent protein and 20 percent fat. At two weeks of age, provide fresh water at all times. At this time you can also offer some fine hay and a small amount of grain. Wean kids when grain intake reaches 1/4 pound and hay is being eaten regularly.

Sharing the Dairy

My favorite option is to allow the kids to stay with their mom for three to four weeks and then separate them from dusk to dawn. I will only do this if I have at least two kids close in age. They'll bleat for a short while then take solace and warmth in each other and settle down for the night. At first light, I head back into the barn to gather my milk and reunite doe and kids for the remainder of the day.

The kids immediately get enough milk to satisfy their overnight hunger, and within an hour, you'll see them filling up their bellies completely. You should not remove a kid from his dam if he is underweight or weak, or if your barn is unseasonably cold.

By the start of the second week, kids will start eating a bit of hay or nibbling pasture. As soon as they do, you can introduce a kid-ration to their diets.

Share the dairy! These nearly weaned kids have been off their dam's milk for evening feedings since they were only a month old. Separating them for the evening, I'd milk the doe once per day and let the kids have the rest once they were chewing on some solid food.

Kids who will be vaccinated usually get enterotoxemia and tetanus shots at eight weeks of age, followed by a booster four weeks later.
At this time, the dairy breed kid could also be weaned.

If you are cautious or concerned about this weaning age and are
calling the veterinarian in for first vaccinations, you could get a sec-

LEFT The meat breeds, like this Boer dam and kid, will self-manage the process of weaning. No forced separation or crying is required.

ond opinion on the condition of each kid. The rule of thumb for weaning is to do so when the kid has tripled his birth weight.

Safe and complete separation from the doe is the only way to effectively wean a kid, and yes, you can expect some loud wailing and calling from both doe and kid for the next two days.

Meat breeds do not need to be weaned—they'll naturally progress to this state on their own. Angora Kids that have not naturally weaned should be separated at least two months before their dams are rebred.

You can bottle-feed and wean the meat and fiber breeds to suit your personal farm strategy. If fiber kids will be kept as wethers, they will be friendlier to humans when started as bottle babies, but remember the level of dedication and the amount of time required to bottle feed. Meat breed doelings planned for future reproduction should be left with their mother and allowed to self-wean. Meat breed bucklings should be weaned by four months and separated from does and doelings well before they reach puberty if not castrated.

Keeping Doe and Kid Records

If you will be keeping your does for repeated breeding over the years, record each one's length of gestation in days and the hour she began labor. She will be a sure bet to follow the exact same schedule next year.

Start your barn records off right before you get too busy training, bonding and playing with young kids. I use and keep data on forms which you can download from www.KeepingFarmAnimals.com, but I also like to keep a notebook for each kid with birth and weekly weight, observations during kidding and feeding, and interactions between siblings and doe.

QUICK REFERENCE BREEDING INFORMATION

Breeding Criteria		
Female		
Puberty		7–10 months
Minimum Weight to Breed		70–75% of estimated adult weight
Estrous cycle		
	Length	18–22 days
	Duration	12–36 hours
	Signs	Bleating, wagging, pacing, mounting
Ovulation		12–36 hours from start of standing heat
Gestation		145–155 days
Breeding Season		Starting when days shorten, lasting approximately 5 months
Male		
Puberty Age		4–8 months
Minimum Age to Breed		8+ months onward
Breeding Season		Year round
Breeding Ratio		1 buck per 20–30 does

11

Goat Grooming
Practices

A HOOF TRIM is one maintenance task that doesn't take much time, but is an absolute necessity. Without proper hoof care, your goat can become sick, lame, or permanently crippled. Frequent attention to the matter, no matter how minor, is less time consuming and far easier than infrequent major trimming.

From the age of two months, I make all goats stand and allow inspection of, or a slight trim of, their hooves every month—no exceptions. Doing so at such a tender age helps them get used to the feel of the shears and file and establishes who is the boss in our relationship.

If you pasture dairy goats on somewhat rocky pasture or in a field that has centuries-old rock fences, the hoof will stay trim a little longer. Goats raised on a small yard could be given a small concrete slab or a stack of large rocks to climb on to help wear the hoof wall down. These aids may minimize the amount to be trimmed but will not excuse the monthly trim.

Our meat herd spent a substantial amount of time on rocks and rough terrain, and there were always a few with overgrown hooves by the month's end. Hooves, like hair, grow at different rates on different goats—they all need to be checked regularly.

To Trim a Hoof

Confine each goat in a manner that prevents escape. A dairy stand with stanchions, a leash and collar tied to a barn wall or fence, or having a helpful friend with you are all great ways to keep a goat still for the amount of time it takes to trim hooves.

If your goat is being particularly difficult, gently lean your shoulder into her side with her other side against a wall while you work. Angoras and kids can be "sat" on their rumps, leaning their backs against your legs while you work.

Stay calm while you work, and you will find that your goat remains calm as well.

1. Soft hooves are easier to trim than dry hooves. An hour in a yard with morning dew will expedite the task.

2. Bring shears, a rasp, and hydrogen peroxide to the barn with you.

3. Wear gloves.

4. Constrain the goat in a manner that allows you to easily and comfortably bend her leg for observation of, and work on, the entire bottom of the hoof.

The Anatomy of a Hoof

Step 1: Gently scrape out any impacted dirt in the curve of the hoof with a hoof pick or the point of closed shears. You are working gently inside the wall. Impacted dirt at the toe of the hoof does not need to be forced out. Get the debris that falls out easily as you trim. It will all have fallen away by the time the hoof has been fully trimmed.

Step 2: Snip away the overgrowth at the toe of the hoof (it will be the longest), then move onto the sides and remove any folds or excess found there. Your objective is a level foot—a hoof wall that just barely extends and protects the frog. The center area of the hoof, the inside wall, can also be trimmed and should be marginally smaller than the outer wall. This ensures that the weight of the goat is on the outer edges of the hoof.

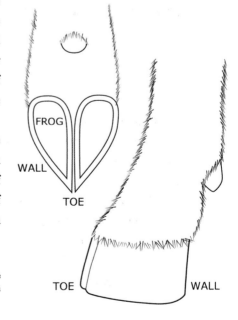

Step 3: Use a hoof rasp to create a smooth finish on the hoof wall. You can also shave or plane the bottom of the foot if it is overgrown with a hoof knife until you begin to see a pinkish tone on the base of the inner hoof.

RIGHT: A nicely trimmed foot. The base of the hoof is parallel to the Coronary Band (the area where the hair ends and the cuticle begins).

Kids and adults alike love to climb on rocks. Doing so also helps to keep their hooves trim between grooming sessions.

Although you would only purchase healthy and well cared for goats, there may be times when a neglected goat with a severely overgrown hoof lands on your doorstep, and you have to nurse her back to health. In this situation, while you're quarantining the goat, trim her hoof every five or six days, taking only a little of the excess off each time. It may take the full quarantine month, but you will have a neat and trim hoof with minimal stress.

Should you overtrim a hoof to the point of drawing blood, clean the wound with hydrogen peroxide or iodine and check daily for any signs of infection. Usually the moment the goat steps down on the hoof, the blood stops. Should the cut be anything more than superficial, stop the bleeding with some blood-stop powder and keep the goat in a clean environment until the cut has healed. Consult your barn records regarding last tetanus vaccination and with your veterinarian regarding antibiotics if you're concerned about a hoof infection after a few days.

Castration

Having consulted with hundreds of people by e-mail about the cute little buckling they brought home from a farm or auction and heard every excuse imaginable on why these new owners didn't want to castrate the young goat, I've nearly given up replying. In all but a few months, the essence of the emails turn from "cute and fun" to "stinky and how do I get rid of him?"

If you are keen on growing a buckling as a pet, for meat, or for profit, please castrate him unless you have a dedicated buyer lined up for a future sale.

Castrated goats are called wethers and the act of castration, wethering.

Bucklings, left to mature without being wethered, will only be good for breeding—a job that should be left to the best of bucks.

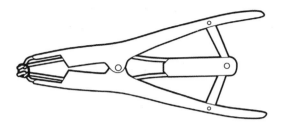

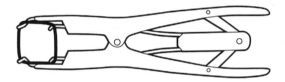

The case for castration is strong. Meat from a grown buck is distasteful to the majority of North Americans, and wethers will grow faster, make better pets, and don't need to be separated from fertile does.

This task can be put off until anytime before puberty, but it is less risky and easier on the buckling if performed early in his youth, preferably before weaning. When the buckling is three to four weeks old, and with the help of a lamb Elastrator, the process is quick and pain free. The exception here is the buckling you plan to keep as a wether. Wait as long as you can to castrate him to prevent the risk of Urinary Calculi.

Elastrators are available at your feed supply store. The process is no more involved than taking a strong set of pliers to stretch a heavy elastic band around the scrotum of the buckling. Within a few days, the testicles wither, die, and fall off. There are no known risks or complications that I am aware of.

Instructions are on the Elastrator box, but most feed supply staff can demonstrate for you in the store. Having a friend who keeps goats or sheep might also come in handy if you're nervous to try it on your own the first time.

You could call in the veterinarian or pay an experienced local to perform the task, but the third-party cost nullifies the economics of growing your own food.

Dis-Budding Kids

You secure the best possible future for dairy kids born on your farm by dis-budding them. You may have learned to work around the horns or never had a problem with horned goats, but the general population certainly does. Whether I will keep or sell a goat, I know it will have a higher value to any farm if de-horned.

As a safety measure, kids should have a tetanus immunity before a dis-budding procedure is performed.

- If the kid remained with his mother, drinking colostrum from birth, and she had been vaccinated for tetanus in the two months prior to kidding, enough passive immunity exists for the first four weeks of life.
- If the kid is over four weeks of age, if he hasn't consumed colostrum, or if his dam wasn't vaccinated late in gestation, a tetanus vaccination can be given during dis-budding. Standard pre-scribed rate for kids is 250–300 IU of antitoxin.

The technique of dis-budding is to kill off existing horn matter which prevents further growth. This can be done by caustic paste, banding, or with a dis-budding iron.

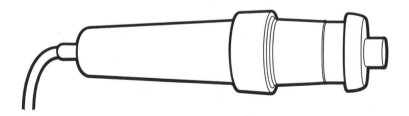

I am not certain what bothers me more about the dis-budding process—the bleating of the kids or knowing that I'm the one doling out the pain to such a young animal. It helps to know that young kids will bleat when held against their will anyway and that their anguish will be short lived. I've seen kids less than two minutes after dis-budding, butting heads with other kids; and young does just moments after having a band applied, pushing their way to the hay feeder.

Caustic Paste Removal

Caustic paste is the old-timers' method of removing buds. This substance is still being sold for calf dis-budding but can cause blindness if applied incorrectly. The paste burns through hair and skin and in my opinion is better left on the shelf of the feed supply store. Skilled herders trim the hair around the bud, apply petroleum jelly to the surrounding area, and then dab on the caustic.

Repeated applications of caustic paste may need to be added over the next few days, and each application holds the opportunity for something to go wrong. When used on kids a helper would need to hold each kid for at least ½ hour—while they bleat and cry and squirm to get away and back to momma. Without confinement, the paste could shake off and burn other areas of the kid's face, the dam's udder, other animals and barn walls.

Banding Horns

Banding is most often reserved for goats with established horns, although it can be performed on kids with small buds. In concept it is much like castration. Strong rubber bands are placed at the base of the horn until the tissue dies and the horn falls off.

To be effective, the band, once in place, should not move from its position (a tall order given the shape of a horn). While most people tape the band in place, others immobilize the head, give a local anesthetic, and make a superficial scalpel incision just below the bottom of the horn ridge. The band is then rolled onto the horn, resting within the incision, and the area sprayed with an antiseptic. Nubs and burrs are said to be less common with this method over the taped method.

Dis-Budding Iron

The electric dis-budding iron is available at most farm supply stores. To help ease the pain and potential for complications, trim the hair around the bud and feed each kid a baby aspirin a few minutes before the procedure. Once the iron is hot enough to mark a small piece of wood, it is held to each horn bud a few seconds at a time to mark the spot and then alternatively for fifteen to thirty seconds until the skull bone is felt under the iron. At this point, the bud can be scooped and removed with a forward motion, and the skull bone will be exposed.

BELOW: This young Saanen goat has been left too long for the dis-budding iron. She will either have to grow up with horns or have them removed by banding.

The iron also is used to cauterize any spots that are bleeding on the kid's head at this time. Each area then gets an application of antiseptic (liquid, powder, or spray on) to minimize chance of infection. You could hold a bag of crushed ice on each bud for thirty seconds to numb the pain, but I haven't found it made much difference. In fact, it only delays the reunion of kid to momma and the favored consolation—a nudge at the udder. If your goat is bottle-fed, have a bottle on hand for an instant distraction.

(12)

Appreciating Goat Milk

IF YOU HAVEN'T tried goat's milk yet, or if you've tried some and been unimpressed, I invite you to try it again—from a reputable dairy, filtered, fresh and ice cold.

Fresh milk from a doe that has had proper care tastes so similar to the grocer's version of homogenized milk that only the most sophisticated palate can tell the difference. That same sophisticated palate will come to prefer the goat's milk for its rich, smooth flavor if given the opportunity to repeatedly savor truly fresh and wholesome goat's milk. How do I know? Because I'm that sophisticated palate, or at least I like to think I am.

As a child growing up in Massachusetts, I often spent afternoons and weekends at a friend's house. Her father ran the dairy up the road and brought home fresh milk every day after work. On the whole, I didn't like milk. In fact, I disliked it so much that I never ate cheese and seldom ate ice cream.

From what I can remember, I'd never been sick from milk, I just didn't like it. Unless I was at Nancy's house. The first glass I drank to be respectful, it had been served to me at dinner. The second glass was by choice. This wasn't the milk I was being served at home but I didn't know enough to ask about the difference. I think it must have been raw milk.

Forty years later, I would never think or ordering milk in a restaurant or using cow milk for cooking at home, but I have no trouble consuming cheese, butter, and glasses of fresh raw goat milk from the farm.

COMPARISON OF MILK

	Goat	Cow	Human
Fat %	3.8	3.6	1.1
Protein %	3	3	4
Calories per 100 ml	67	69	68
Vitamin A (i.u. G fat)	39	21	32
Vitamin B (UG / 100 ml)	68	45	17
Vitamin C (mg / 100 ml)	2	2	3
Vitamin D (i.u. G fat)	0.7	0.7	0.3
Riboflavin (ug / 100 ml)	210	159	26
Calcium	0.19	0.18	0.04
Cholesterol (mg / 100 ml)	12	15	20
Iron	0.07	0.06	0.2
Phosphorus	0.27	0.23	0.06

Nearly 70 percent of the world's population can't be wrong (that's how many people drink goats' milk regularly). In fact I think that North America may be the only continent that drinks milk from a cow as standard practice. Milk that has been smashed (the homogenization process), heated to the point of cooking all the goodness out of it (pasteurization), enhanced (with added vitamins and dry milk solids), blended (with many other cows' milk) and stored for transportation.

Even though I'm passionate about all the reasons not to drink cow's milk, they aren't the reasons that I drink goat's milk. I drink it because it simply tastes great to me, is healthier for my body, and it's far easier on my digestion.

Naturally Homogenized

Goat's milk is naturally homogenized. There are both positive and negative aspects to this. The positive aspect is that it is one step closer to consumption direct from the source. I appreciate that. Lately I've had an aversion to all the ways we touch, process, extract, add to, and mess with

Goat Milk Tastes Bad

I hear it all the time. Someone gave you a glass or quart of goat's milk, and you tried and disliked it. But I want to challenge you to try it one more time, fresh from a dairy, and ice cold because if you didn't like it, you must have had some bad milk.

There are many ways the taste of milk can be tainted or found as pungent. Milk from a doe kept within two hundred feet of a buck will taste off; or if the doe hasn't been fed the right foods. Milk is extremely delicate and absorbent. It is a direct by-product of the food consumed by a doe and the air at collection. Good grasses and hay, a little grain, and lots of fresh air will make delicious milk. Does housed with mature bucks and existing on pine needles and garlic chives will not.

natural sources. With so many changes (many on a molecular level) to the food we consume, it really isn't surprising that so many of us are ill.

Naturally homogenized milk means that it won't require a smashing of the fat cells within the milk to make it stable. That is precisely what we do to cow's milk to homogenize it, alter the cells

BELOW: No boys allowed. These Saanen and Sable does are kept separate from bucks.

within so that we won't be inconvenienced by a separation of butterfat to milk.

Which brings me to the downside of natural homogenization. Separating the buttermilk content in goat's milk to make fresh, all natural butter is a little more work. It can be done and it is truly delicious, but without a separator it is a little more laborious.

High Butterfat Content Makes the Best Cheese

Many of the dairy breeds produce milk that has a butterfat content of 6 percent or more. This feature alone makes goat's milk a natural choice for creating delicious homemade cheese. The simplest goat cheese can be made in under twenty minutes with less than two days of curing using only milk, salt, and a little lemon juice.

Butterfat content will differ from one doe to another, between stages of lactation, and based on the food they're consuming. In the spring and summer when does are on lush green pasture, they will produce more milk, but it will be lower in butterfat. When milk yield is low, such as the end of their lactation cycle; and when dry matter is the primary food, fat and protein content is higher. This it the best time of year to make some delicious cheeses for your family.

Getting to Know the Dairy Doe

In the dairy industry, goat milk is never measured in gallons or quarts but by weight. Since my goal in keeping dairy goats is to have fresh milk and cheese for our family's table, I don't weigh my milk or attempt to gain recognition from the various associations for butterfat content or quantity.

This may be something that interests you if you plan on starting a small dairy or raising-to-sell top-quality dairy does. You can get started with the same routine for milk collection and record keeping as I do; you'll just need a small scale in the barn for weighing each doe's output.

I record both morning and evening output for each doe in a little notebook hanging from a hook in the goat shed in ounces. The only reason I keep or have ever used these records is when I need some insight into the health of a doe and when I think that I might need to adjust a doe's feed quantity.

A well-managed, full-sized, dairy doe is going to supply you with eight hundred to nine hundred quarts of milk over the course of a year. That's about a gallon per day. The smaller breeds, crossbreeds, or grades (of unknown breed) will supply less. If your family won't drink a gallon of milk per day (multiplied by two because you should really have more than one goat), you can find a smaller or lesser doe or make cheese and butter. You can make a pound of cheese out of a gallon of milk.

Most dairy does produce milk for ten months of the year and require breeding to supply milk for the following year. Some does will milk right through a second year without being bred. If the doe you own produces a low quantity or if she dries out earlier than the expected ten months, she is still a great milker for a small family. The way I see it, for what they cost us in feed and time, even a quart of milk per day for two hundred days of the year, is ample return.

Most does are at the height of their milk production in their fourth and fifth years.

Raw Milk vs. Pasteurized

Many books, newspaper articles, research papers, and more have been written on this topic. I urge you to do your own research on the matter for a better understanding of the controversy before making any changes to your life choices.

Here are the facts as I understand them and the choices I've made based on that understanding.

The process of pasteurization kills bacteria, unhealthy organisms, and goat-to-human communicable diseases; the most well known being tuberculosis. Unfortunately, pasteurization also nullifies the healthiest aspects of raw milk.

After careful contemplation and research, I drink and make cheese from raw milk. I don't propose that my choices are in your best interest; that is something you will have to decide for yourself. Should you decide to forgo pasteurization, please have your dairy goats tested by a veterinarian when they enter your farm, keep them quarantined until all tests have returned from the vet's office, vaccinate each one annually, and stay on top of current regional news for wildlife alerts and disease breakouts among livestock. Collect and store all milk in the most careful manner and date every container of milk that enters your fridge.

One of the primary reasons pasteurization is required in the commercial dairy industry is tuberculosis. I have been told that it is a "last century" disease, all but wiped out. I have also been told that goats are immune to tuberculosis. Neither are correct. In Carmarthenshire, United Kingdom (2008), an outbreak of tuberculosis among a disbanded herd received a lot of media attention. Goats can certainly contract tuberculosis, both from infected cattle and from wildlife, and can pass the disease onto their keepers.

The Making of Milk

A goat's milk production is dependent on the kidding cycle as well as your milking schedule. Peak production is two months after kidding, but good dairy goats continue to produce milk long after kids are weaned.

When a doe first gives birth, her body begins producing milk. The term for this is freshening. For the next eight weeks, her supply increases, levels off for four months, and then supply slowly decreases until she kids again.

Standard small-herd practice is to breed the milkers once per year, dry them out for two months previous to kidding, allow them to keep their kids around the clock for at least two weeks, and then split the milk supply with the kids until they are weaned. After weaning, all of the doe's milk will be available for your personal use until she is dried off again, a few months after rebreeding. Under these conditions, does supply milk for 240–300 days of the year.

I also mentioned that quantity of milk is dependent on your milking sched-

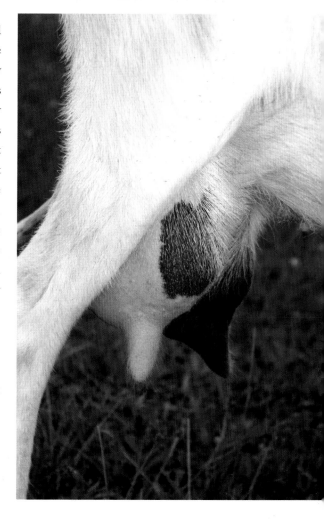

RIGHT: The udder of a young doe is beginning to get a little full. Within an hour, this doe will be at the door, waiting to be milked.

ule. Goats are creatures of habit and love life's rituals and routines. Decide early on in your doe's keep, how often you will milk her, and the approximate time of day.

Most does are born predisposed to, or raised for, two milking sessions per day: once early in the morning and once at dusk. You can alter an already established routine slowly without loss of supply. You can also dry off a doe early if you are planning a vacation and don't have friends, family, or neighbors who are willing to take on the task of twice-daily milking.

You could also increase or decrease the number of times per day that you milk a doe to get more or less milk from her as your family requires. Keep in mind, however, that once you opt to milk three times per day (for maximum quantity) or to raise a doe for once a day milking (for less quantity), you could be altering her habits, her comfort and her production quantity for the remainder of her life. Should you ever sell her, these decisions will also affect her new owner.

Negative factors affecting milk production are stress, insufficient natural or artificial lighting, weather, injury, feed quality, illness, or major disruptions in their routines.

Dairy does thrive on routine. You get to set the time that she can expect you to milk her, but for best results, remain consistent with her schedule and keep milking sessions evenly spaced (i.e., twice daily, twelve hours apart).

Equipment and Supplies For Milking Goats

Although you could get by in a pinch with a high-quality stainless steel bowl from your kitchen, a cheap milk strainer, and some quart canning jars, if you will be milking twice a day for two hundred or more days per year, you might as well have the proper equipment.

- Milk stand (with or without stanchions)
- Clippers and a brush to trim and brush hair away from back, sides and udders
- Udder wash to remove any debris and to sanitize hands
- Small bowl or mug used as a strip cup (the first milk from each teat is squirted in here for a visual quality check)
- Stainless steel milk pail
- A made-for-goats teat dip to prevent bacterial infections
- Bag balm (to prevent chafing or chapping of the udder and teats)
- Notebook to record milk amount from every session
- Dairy strainer with disposable milk filters
- Milk scale for weighing milk per session (I don't personally use these.)
- Home pasteurizer
- Glass milk jars

BELOW: Milking stands come in a wide variety of shapes, sizes, and composition, but they should be a minimum 18 inches wide and 3 1/2 feet in length. You can find them at goat supply stores, in the local classifieds, or commission a carpenter to make them for you. Stanchions to hold the goat's head in place aren't necessary but do hold the goat in place until you have finished milking and caring for her udder.

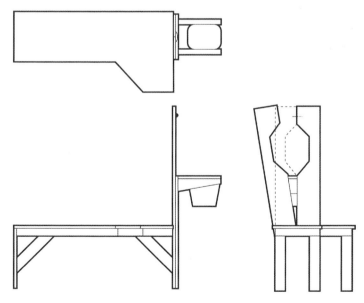

- Chlorine bleach
- Stiff bristle (plastic) brush
- Dairy acid cleaner

How to Milk a Goat (or Not)

When it comes right down to brass tacks, farmers have been milking goats for centuries without all the fancy equipment. Half embarrassed to tell you my own story but wanting to make a point about getting by with what you have, what follows in the next three paragraphs is a descriptive account of my first adventure in milking a goat.

A doe and her soon-to-be-weaned kid landed in my barn, and I, knowing nothing other than the rules of sanitation, proceeded to milk her twice daily.

She was young and small and in retrospect, I believe she had been bred too young. This worked in our favor as she had never given milk and I had never taken it. She would not stand idly on the barn floor and let me milk her. I had never heard of a stanchion, but I was determined that she stand without fussing or kicking over the bucket.

The only way I could hold the doe in place on my own was to lock her head between my legs, bend forward across her back with my head off to one side, and reach under to milk her. I did this twice a day, every day, for two weeks before I started to seek alternatives.

Today, I shudder at the thought of all the ways this manner of milking might have gone wrong. She was not elevated, which meant that all too easily, she could have kicked debris into my milk bucket. I simply let her go after milking—had she laid down or her teats come into contact with bacteria, she could have easily developed mastitis. Yes, we drank healthy delicious milk every day for those first few weeks, but thankfully today, I am a better steward of my animals.

The Better Way to Milk a Goat

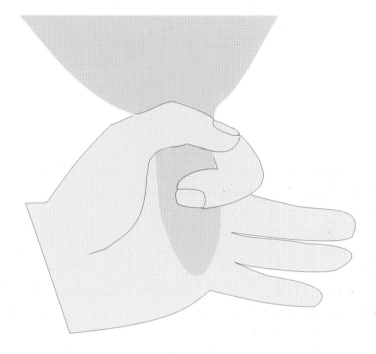

SANITIZING EQUIPMENT

Sanitize all equipment before and after milking, including udder, teats, and hands to ensure a healthy happy doe and high-quality milk.

To sanitize equipment between uses, first rinse in warm water to remove fat residue. Next scrub with a bristle brush, hot water, dish detergent, and an ounce of chlorine bleach. Rinse in clear water. Never use dishcloths to wash or towels to dry. Allow your equipment to drip dry.

Once a week, wash everything as above but using a dairy acid cleaner in place of dish detergent.

THE MILKING RITUAL

When your doe is accustomed to being milked, she will be ready at the scheduled time and (in my experience) compliant about the process. Have all of your supplies on a nearby shelf then collect your doe.

Quickly brush her back and sides to remove loose hair and hay or anything else that may fall into and contaminate the milk. Once a week or when required, clip any long hairs around the udder. You can do this while she's on the stand, but I prefer to do it beforehand.

Lead her to position. Some owners will give the doe grain at this time with her head secured in stanchions and some will feed grain after milking. If you have a stand, your job will be easier, but if you don't, you can milk her on the floor while you sit on a small stool beside her. I am six feet tall but fairly flexible, and have milked five does consecutively on a little stool without a sore back at the end of the session.

Wash your hands and her udder with an udder wash formulation or warm, slightly soapy water and a clean towel. Dry the doe off with another one-time-use towel.

Perform a strip test in two small cups, one for each teat. This first draw will be the first point of contact for bacteria entering the teat and will also provide an indication if there is anything wrong with your goat or her milk. Look closely for any abnormalities or tints of blood in the milk. If you see anything wrong, continue to milk your doe but don't use the milk.

Position your milking pail directly under the udder, a few inches forward. Lock your forefinger and thumb around a teat at the very base of the udder and draw down. You can milk both at the same time if you are comfortable doing so.

There is a rhythm to milking a doe. You can learn it just by watching the kids. Bump. Lock. Draw.

There will be milk waiting in the teat at milking time, which you will guide out with a squeeze "drawing" downward, not by "pulling" downward. If the difference doesn't make immediate sense to you, think of how you might squeeze all of the toothpaste out of the tube with only one hand.

After a full draw down and into the bucket, release the pressure on your top two fingers to allow more milk to enter the teat, and draw again. Release. Each time that you release your grasp, the teat fills up with milk again.

Many first-time farmers will not get any milk out of the teat on their first try. Don't worry if this happens to you. The cause is most often inadequate pressure at the top of the teat and the milk moves back up and into the udder instead of along the teat and into the bucket. Squeeze just a little tighter on top and try again. Never squeeze her udder, only her teat and only as hard as is warranted to help the milk flow.

A few tries and you will have it. Your doe relaxes and you can move onto learning how to milk her with both hands in an alternating left-right-left-right fashion.

When the flow stops, you can give a gentle bump to the udder with your hand still wrapped softly around the teat. Then tighten-squeeze-release once or twice more.

Strain and filter the milk immediately into glass jars and seal with a lid. If you are keeping raw milk and have a small fridge in your goat barn, you can partially immerse the jar into a bucket of cold, ice water waiting in the refrigerator. If you are planning on pasteurizing, cover the jar or container so that minimal light affects it until you get back to the house for pasteurization.

Wipe teats with sanitizing wipes and perform a teat dip to kill any germs and bacteria.

Release the doe to fresh hay and water. She will stand to eat and drink, which is your best defense against teat contamination leading to mastitis until teat orifices have closed again.

Repeat the same process for each doe using clean towels, strip cups, bucket, milk filter, and jars.

What to Do With All that Milk

You can collect, cart, and treat your milk in any manner that suits your style, equipment, and preferences as long as you keep the raw milk out of direct sun or florescent lighting until it has been cooled.

If you will be keeping milk in a raw state, you will want to cool it off as quickly as possible after collection. The rule of thumb is to cool the milk from 101 degrees Fahrenheit (at time of collection) down to 38 degrees Fahrenheit within one hour of milking.

In the barn, I filter the milk directly into mason jars then cover the jars getting carted back to the house with a clean towel. Once inside, I cool the milk in the freezer for twenty minutes before transferring into the fridge.

As water will cool off jarred liquids faster than cool air around the liquid, you could put full jars in a sink or washtub full of ice and water. An eco-friendly option to ice cubes are those blue reusable freezer packs. Many goat keepers will filter directly into stainless jugs at the barn and cool the milk, still in the jug, by immersion into an large ice-filled pail.

Pasteurize milk on your stove top or in a specialty countertop pasteurization machine. The machines are currently selling between $400–500 for a brand-new model. If you're on a budget, you might be able to find a gently used model on ebay for much less.

To pasteurize without a machine, you'll only need a double boiler and a liquid thermometer. There are two types of thermometers that are appropriate—a floating one and one that snaps onto the side of the pot.

Stir continuously while heating your fresh filtered milk to 165 degrees Fahrenheit and hold the temperature there for fifteen seconds. Immediately transfer the top half of the double boiler to a sink of ice and water stirring continuously until the milk cools to 40 degrees Fahrenheit. Transfer the milk to storage jars before placing in the refrigerator for immediate use or the freezer for later use.

Excess milk can be frozen in glass containers. Thaw frozen milk in the refrigerator for twenty-four hours and shake for a minute before using. If you notice a change in flavor after freezing, use the milk for baking or cheese, butter, and soap making.

I often bring milk in, chill it, and immediately begin making cheese. As I'm usually pressed for time and cheese is quickly eaten up in our house, I use a simple recipe that makes a soft, fresh cheese in about two weeks. My favorite, as well as more involved recipes, are in the back of the book. Don't stop at the included recipes though. There is a rising interest in homemade cheeses, and you'll find a wide assortment of published books on the market for further experimentation and study.

Three Simple Milk Rules

1. Never add new milk to old milk or warm milk to cooled milk.
2. Pasteurizing or not, chill milk within one hour of collection.
3. Milk should never be exposed to sunlight or florescent light before and during the cooling process.

13
Fiber Goats and Production Specifics

THE CARE, BREEDING practice and profit from Angora and cashmere-producing goats has very few differences from that of the other breeds unless you plan on owning a large herd for farming income. If that is your goal, specialized knowledge and management practices should be investigated for the highest returns and chance of success.

Cashmere Production and Collection.

The downlike hairs that grow in small clusters under the main coat of a goat is cashmere. As stated in the breed section, many breeds are capable of producing cashmere, but not many produce enough quantity or are of fine enough quality to make the collection worthy of time and energy.

Goats in the top in their class for producing cashmere sell for hundreds of dollars or more at the time of writing. They are sold as breeding stock for farmers already involved in or interested in entering the cashmere production business. These goats may only produce 1/4 to 1/3 of a pound of cashmere per year, but due to their large size, the sale of meat is a viable secondary market. Over the last fifteen years, I have seen cashmere prices fluctuate between $40–$100 per pound.

Cashmere fiber is commonly collected by combing the goat daily and collecting the fiber exclusively, but some owners shear their herds, netting just over two pounds per goat of a mixed-grade fiber (approximately 40 percent cashmere). Daily combing for fifteen to twenty consecutive days right before winter solstice when they naturally begin to shed the fiber nets the highest yield.

Save collected fibers in a cardboard box or paper bag without the long primary hairs that your comb picks up. Expect to collect anywhere between 1/3 to 1 full pound per goat.

Angora

Without being repetitive to other information in this book, here are the differences of care and production for Angora goats:

- Angoras can annually produce 25 percent of their own body weight in fiber with an average production of ten to sixteen pounds.
- Allow one buck per fifty does if pasture breeding. One buck can service 150 does (over the span of sixty days) if visitation breeding is used.
- Angoras are less interested in busting out of or jumping over fences than other breeds.
- Newborn kids weigh four to six pounds. By three to five weeks, fine mohair fibers appear between course newborn hairs.
- Does nursing a kid past the age of five months, and being bred again in the spring will produce a low quality fleece. Wean kids early enough to give the doe two months of rest before her next breeding.

The Angora goat creates enough fiber to make collection viable through shearing. Color ranges are from white, pale silver, and tan to cinnamon red, blue, and black. An angora goat's short fibers are mohair and have the most value. The longer fibers are called kemp.

Yearlings give the finest hair, therefore yielding the highest price. Kid hair is used for clothing with a luxurious softness unmatched by any other fiber. Shearing isn't a right of passage based on age. Any kid with hair four inches or longer can be sheared with the rest of the herd, but kid hair will be collected first on shearing day and sold later in separate batches.

Adult Angora hair is used mostly in fine carpets and upholstery. The average doe, buck or wether hair is 4 to 6 inches in length at shearing. Angora fiber grows between 3/4" to 1 inch per month.

Expect yields of three to four pounds from a yearling and six to twelve pounds per shearing from an adult Angora. Fleeces are worth

The fiber from first-shear kids are the finest and net the highest price.

considerably less when vegetation, lice, or nits are present. Mohair most coveted by hand-spinners is soft and full of curly ringlets that are evenly crimped from tip to tip. Finer crimp will net the highest prices.

At the time of writing, individual bucks and does sold as breeding stock start at $300.

Shearing the Angora Goat

Shearing of Angoras is performed just before kidding (early spring) and then again in the fall (between weaning and breeding for the next season). Carefully controlled and calculated breeding is based on seasonal projected temperatures for the following spring. Shearing too early in the spring or too late in the fall could leave an Angora without a suitable coat in cold weather. Angoras need ample time to regrow their coats before the weather turns cold.

You could shear your goats yourself or hire a professional to either (a) train you on technique, or (b) to shear your goats for you. A few days before shearing, clean out the barn, bring the goats in, and keep them dry. Wet mohair is not only difficult to remove, it is also prone to mold during storage. You do not need to wash and dry the fleece before sale, but you could net a higher price if you do.

Shear kids first on a clean workspace. Have a helper on hand to pick short hairs and any irregular, stained, or matted clumps out of the fleece. Each fleece is then rolled inside out and stored individually in a cloth or paper bag, or cardboard box. Tag each package by the goat's name so you can track fleece weights from season to season. When you are finished with the kids, sweep the area clean and follow the same procedure with adult Angoras.

Wash fleece in hot water and gentle detergent, leaving to soak for thirty minutes or more. Rinse well with hot water and soak/wash again if necessary. A few drops of vinegar will bring out a beautiful shine. Fleece can be dried flat on towels or a screen. Drying may take several days, but you can speed up the process with a fan or blow dryer on a warm setting.

After shearing, you may need to keep your goats inside or wearing a goat coat for up to thirty days (the average length of time to grow 3/4" to 1" of hair) if the weather is unseasonably harsh. Coats are inexpensive in comparison to complications of exposure such as sunburn, hypothermia, or pneumonia.

ABOVE: The Angora goat comes in many colors, but the most common are white. They are slower to grow than the other breeds, not reaching their full weight until two years of age.

Favorite Recipes

Goat Cheese

HOMEMADE CHEESE IS one of the true joys of keeping dairy goats. With very few ingredients and supplies, you can make delicious and healthy cheeses that your family will enjoy. In most cases, all you will need are stainless steel pots, a colander, a thermometer, and cheesecloth—but many fun and innovative supplies are becoming more readily available from cheese making supply stores and websites.

There are just a few items I need to share with you about basic cheese making before you begin.

Pasteurization of milk alters the molecular structure of milk solids, so if you don't like the finished cheese from any of these recipes, try switching to raw milk as the base.

When a recipe calls for added ingredients, buy and use an organic version for a truer taste, healthier alternative, and finer texture. The same premise stands for added water—use filtered or spring water over chlorinated or tap water.

The basic theory of cheese is to curdle the milk and work with the curds toward the end product. There are as many ways to do this as there are types of cheese. One of the traditional methods is to add an enzyme called rennet. Rennet comes from a number of sources but in European countries (where most cheeses originate), it is extracted from the fourth stomach of non-weaned calves—the secondary sub-market of veal. In North America, we use a plant-based extract to manufacture rennet but 90 percent of the market share is the product of a genetically modified plant. Environmentally or socially conscious readers should perform their own due diligence based on their comfort level of these issues.

Fast Soft Goat Cheese

| 1 gallon goat milk | 1 tablespoon salt |
| ½ cup apple cider vinegar | |

After chilling the milk, warm it on the stove to 175 degrees Fahrenheit. Stir gently for two to three minutes and remove from heat.

Add vinegar and continue to stir gently until soft curds form.

Pour into a colander lined with cheesecloth resting in a stainless steel bowl.

Let rest for two minutes and then lift the cheesecloth with the curds above the colander. Gently squeeze to remove excess whey.

Empty the curds into a glass or stainless steel bowl, add salt, and mix together with clean hands.

Form the curds into a ball or log, wrap in plastic wrap and store in the refrigerator for at least two days before eating. If you'll be entertaining or giving your cheese to friends, try rolling the log in cracked black peppercorns or a blend of your favorite herbs and spices before serving.

Goat Milk Feta Cheese

Feta is a deliciously salty and tangy cheese. It is excellent in Mediterranean salads, crumbled on top of pizza, and added to the top of bruschetta. The longer you cure the feta, the more salt it will absorb. You can rinse your cured feta in filtered water before using if you find it too salty for your tastes, but don't skimp on the salt when making it. Start this recipe three to four days before you need feta.

I gallon fresh, chilled goat milk	½ tablet rennet dissolved in ¼ cup filtered water
I tablespoon fresh yogurt (goat or Greek unflavored yogurt is best)	salt

In a double boiler, warm milk to 80 degrees fahrenheit while stirring slowly.

Remove from heat and add yogurt slowly. Allow the mixture to stand covered on the kitchen counter for one hour.

Dissolve the rennet in the water and stir gently into the milk-yogurt mixture. Allow the mixture to stand covered on the kitchen counter for twelve hours.

The curds will appear as a solid mass. Cut into cubes, approximately ½ inch in size. Stir gently and let the curds rest for twenty minutes.

Strain the mixture through cheesecloth, slightly elevated over a large bowl for at least two hours, reserving the whey. Pour the whey into glass jars and refrigerate.

Empty the curds into a stainless or glass bowl and thoroughly mix in the salt. Doing so will break up the curds.

Transfer the curds to a fresh cheesecloth lined colander or cheese mold, folding the cheesecloth over the top to cover. To help remaining whey to drain, add a weight to the top of the cheesecloth. Refrigerate for one full day.

Make a pickling brine of the reserved whey plus salt. Three to four cups of whey to four to five tablespoons of salt. You may want to alter the ratio based on your personal taste for future recipes, but don't skimp too much on the salt, or you will not achieve the full feta taste.

Remove the cheese from the cheesecloth, cut into one and a half inch cubes, and transfer to a wide mouth mason jar or glass bowl. Add brine to fully cover the cheese. Cover and refrigerate for a minimum of three days. Use within two weeks.

Calzone with Goat Cheese

Makes enough dough for four calzones.

If you haven't tried calzones before, they are somewhat like a pan-zerotti, only they're baked instead of fried. You can fill these with two ounces each of your favorite tomato sauce and anything else you like. My favorite version contains two ounces of soft goat cheese, two ounces of diced, roasted red pepper, and two ounces of cooked chicken breast.

1 cup lukewarm water	1 teaspoon salt
½ teaspoon sugar	3 cups flour, plus extra for dusting
2 ¼ teaspoons active dry yeast (or one small packet)	1 tablespoon extra-virgin olive oil

Combine water, yeast, salt and sugar, and stir to dissolve. Set aside for fifteen minutes allowing the yeast to activate.

Add flour and mix until smooth.

Turn the dough onto a floured work surface and knead for ten minutes or until the dough feels smooth and elastic.

Lightly oil a bowl and place dough inside to rest for one hour or until doubled in size.

Quarter the dough, shaping each piece into a ball and roll flat into nine-inch circles.

Place your round onto a cookie sheet, add you favorite ingredients in the center, and fold up both sides. Seal the edges by crimping with your fingers or a fork, leaving a small area at the top of the calzone for air to escape.

Brush the entire calzone with an egg wash or olive oil, whichever you prefer. (Olive oil makes a slightly crunchy calzone wrapper).

Bake until golden in a preheated oven at 400 degrees Fahrenheit for eighteen to twenty minutes.

Curried Goat

A traditional Caribbean dish. Serves 4.

1 ½ pounds diced goat meat	4 chopped tomatoes
3 tbsp curry powder	2 diced onions
1 tsp fresh or dried thyme	2 crushed cloves of garlic
2 tsp freshly ground black pepper	olive oil
2 tsp coriander seeds, crushed	3 cups water
5 oz vegetable oil	

Season diced goat with curry powder, thyme, black pepper and coriander seeds and store covered in the refrigerator for at least six hours.

Preheat oven to 300 degrees Fahrenheit.

Heat a little olive oil in a pan and cook the goat pieces until golden brown on the outside.

Lightly saute tomatoes, onions, peppers, and garlic in a splash of olive oil and add it to the meat.

Add three cups of water and bring to a boil.

Remove from heat and place the pot in the oven for 1—1.5 hours.

When the goat meat is tender, remove the meat from the pan and simmer the remaining liquid in the pot for twenty minutes.

Spoon over or beside a bed of rice to serve.

Honey and Oatmeal Soap

Have you seen Goat Milk Soap cropping up in drug stores and gift shops lately? If you've tried it, you know that it is one of the gentlest healing soaps on the market.

12 oz lye (Available at the grocery or drug store. Use with extreme caution.)

6 cups of goat milk	⅓ cup honey
5 lbs lard	1 ½ cup finely ground oatmeal (use your blender to grind it up)
½ cup olive oil	

Use stainless steel or glass pots. Use plastic or wooden spoons for stirring (do not plan on using these ever again on anything but soap making).

Slowly add the lye granules to cold goat milk in a large pot; stirring constantly. The milk will darken and may curdle a bit, but just keep stirring. The lye will heat the mixture up. Cool it down to 85–90 degrees Fahrenheit. Add honey.

Heat up the lard and olive oil to match the temperature of the milk/lye mixture.

Slowly add the lard mixture to the lye mixture and stir continuously for fifteen minutes (if you can pick up an electric stick blender from a yard or garage sale for a few dollars, it will save wear and tear on your arm—remember though that it is best not to use it for any purpose other than soap making once you've used it with lye.)

At the end of the fifteen minutes, add in the oatmeal.

Check on and stir thoroughly every ten minutes until a spoonful of the mixture drizzled across the surface leaves a trail (In soap making this is called "trace").

Pour into molds (you can use anything plastic for a mold, even Styrofoam cups). Or put on rubber gloves and shape the slightly air-cooled mixture into soap balls.

If using molds, let the soap set for two days in a cool and airy location of your house. Freeze for about three hours and remove soap from molds.

Your soap will need to cure for three to four weeks in a cool and dry location. Turn the bars or balls every few days to allow air to cure all sides.

Acknowledgments

AS ALWAYS, THE credit for content herein goes to my daughter Veronica. She carried me to this land and worked at my side to explore the joys, hard work, and wonders of raising farm animals and growing our own food. If it were not for her eagerness to try something new and emerge victorious, we would never have brought "Nanny" home from that country auction to nurse back to health. Nanny's lieben (the word for love in Switzerland—the origin of her breed) and appreciation paved the way and opened our hearts for many more goats.

Neither the writing of a book nor the education of its author are solitary events. This book would not have been possible without the co-creation and collaboration of ideals and understanding from a long list of supportive friends, neighbors, and personal resources.

Special thanks go to Michael Foster for being the quintessential country boy in my life. From the time he was just two years old, he has taught me, through actions and deeds, to care for farm animals with an equal balance of compassion and common sense.

Personal thanks for the unending support during the writing of this book go to: Eric Kleinoder, David and Diane Peck, the Bird's Creek Feed and Seed, Ashlie's Book Store, Joy Morren, Chris Elsworth, Brad Burke, Molly Cox, Scott and Lucille Kyle, many of my Facebook friends and fans, and the thousands of subscribers who have e-mailed their own accounts of raising goats since the start of GoodByeCityLife.com

To Ann Treistman, who repeatedly challenged and graciously assisted me in the personal adventure of writing a second book for print publication. Thanks also to the rest of the team at Skyhorse Publishing: Tony Lyons, Bill Wolfsthal, Kristin Kulsavage, and last but not least, designer LeAnna Weller Smith.

Index of Tables

Body Condition Scoring Assessment 19

Goat or Sheep? 22

Average Yield for Full-Size Registered Breeds 28

Quick Facts on Goat Meat 44

Star Milker System Explained 67

Common Complications of Incorrect Nutrition 109

General Ration Guidelines 122

Suggested Meat Goat Breed Rations 123

Common Signs of Internal Parasitic Infestation 129

Protein Content of Food Sources for Goats 123

Dairy Goat Weight Chart 130

Commonly Used Vaccinations 136

Potentially Poisonous Plants 147

Is Your Doe in Heat? 159

Kid Bottle Feeding Schedule 171

Quick Reference Breeding Information 176

Comparison of Milk 192

Three Simple Milk Rules 205

Goat Associations

American Boer Goat Association
1207 S. Bryant Blvd., Suite C
San Angelo, TX 76903
325-486-2242
www.abga.org

American Dairy Goat Association
P O Box 865
Spindale, NC 28160
828-286-3801
www.adga.org

American Kiko Goat Association
82224 Kay Road
Wakita, OK 73771
254-423-5914
www.kikogoats.com

Candian Goat Society
449 Laird Road, Unit 12
Guelph, ON N1G 4W1
519-824-2942
519-824-2534
www.goats.ca

Canadian Livestock Records Corporation
2417 Holly Lane
Ottawa, ON K1V 0M7
613-731-7110
www.clrc.ca

Canadian Meat Goat Association
449 Laird Road, Unit 12
Guelph, ON N1G 4W1
519-824-2942
519-824-2534
canadianmeatgoat.com

International Kiko Goat Association
Registrations and Transfers
c/o Starr Perry
3289 Toccoa Hwy.
Clarkesville, GA 30523
706-969-9759
www.theikga.org

Miniature Dairy Goat Association
PO Box 1534
Woodland, WA 98674
360-225-1938
509-396-9922
www.miniaturedairygoats.net

National Pygmy Goat Association
1932 149th Ave. SE
Snohomish, WA. 98290
425-334-6506
www.npga-pygmy.com

Nigerian Dwarf Goat Association
NDGA Registrar
1927 East 500 North
Ossian, IN 46777
260-307-1984
www.ndga.org

References and Recommended Reading

Getting Your Goat: The Gourmet Guide
Patricia A. Moore
Publisher: Evertype (May 1, 2009)
ISBN-13: 978-1904808251

Goat Medicine
Smith & Sherman
Publisher: Wiley-Blackwell; 2 edition (July 21, 2009)
ISBN-13: 978-0781796439

Home Cheese Making: Recipes for 75 Delicious Cheeses
Publisher: Storey Publishing, LLC; 3 edition (October 14, 2002)
ISBN-13: 978-1580174640

The Merck Veterinary Manual
Cynthia M. Kahn
Publisher: Merck; 10 edition (October 19, 2010)
ISBN-13: 978-0911910933

Milk: The Surprising Story of Milk Through the Ages
Anne Mendelson
Publisher: Knopf (October 7, 2008)
ISBN-13: 978-1400044108

Ruminant Nutrition: Recommended Allowances and Feed Tables
R. Jarrige (Editor)
Publisher: John Libbey & Co Ltd (January 1, 1990)
ISBN-13: 978-0861962471

Veterinary Guide for Animal Owners
C. E. Spaulding
Publisher: Rodale Books; Revised edition (February 24, 2001)
ISBN-13: 978-0875964041

Photo and Illustration Credits

Goat Anatomy: Veronica Childs: 18;

Purebred French Alpine Doe, Popular Hill Muscat Champion, Photo Courtesy of Popular Hill Dairy Goat Farm: 39;

Purebred LaMancha Doe, Dancing Diane, Photo Courtesy of Pam Elbert, Agape Oaks: 40;

Purebred Nubian Doe, Poplar Hill Fairfax Ruby, Photo Courtesy of Poplar Hill Dairy Goat Farm: 41;

Purebred Saanen, Violet Vale Age of Discovery, Photo Courtesy of Sara Dzimianski, Violet Vale Dairy Goats: 43;

Purebred Toggenburg Doe, GCH Poplar Hill Boulder Crystal, Photo Courtesy of Poplar Hill Dairy Goat Farm: 44;

Purebred Boer Buck, CSB Ruger Reloaded Ennobled, Photo Courtesy of Circle Star Boers, circlestarboers.com: 46;

Purebred Kiko Buck, Penn Acres Huckleberry, Randy and Cheryl Penn, Penn Acres.com: 47;

Purebred Nigerian Dwarf, Hidden Hollow Wysteria, Michael and Keary Mariannio, Southpaw Acres, southpawacres.net.com: 51;

Purebred African Pygmy, Madaline Mastroianni from Monterey Bay Equestrian Center, montereybayesquestrian.com: 52;

Goat Eating Denim, Thomas de Florie: 96;

Goatlings and Grain, Micghael Richert: 106;
Goat Cheese Calzone, Knut Pettersen: 218;

Rebel Ridge Boer Goats: 159;

Scott Dawson: 185, 186, 199;

Laura Childs: 73, 76, 201;

Unless otherwise indicated, photographs are licensed by Shutterstock.com.

Index

A

Adjusting to new environment, 66
African Pygmy goat, 52
Alpine goat, 38,
American Alpine (see Alpine goat)
American Livestock Breeds Conservancy,
    47, 49
Anglo-Nubian (see Nubian goat)
Angora goat, 34, 50

B

Body condition scoring, 19
Boer goat, 28, 33, 45
British Alpine (see Alpine goat)

C

Chevre, 6
Commercial feed, 103
Common Complications of Incorrect
    Nutrition, 109
Cow's milk, 4
Cross-breeding strategy, 27

D

Dairy goat, 38
Doelings, 34

E

Ear shape 40
    Elf ear, 40
    Gopher ear, 40
Experimental goat, 27
External parasitic infestation, 61

F

Feeding goats, 96
    Components of nutrition, 102
    Digestive system, 104
    Feed consumption, 108
    Feeding style, 100
    Goat hay, 113
    Pasturing, 112
    Ration, grain, supplements, 115
    Variety, 99
    Water, 121
Finding goats, 56
    Bulletin board, 57
    County fairs, 58
    Feed supply store, 56
    Forum newspapers and magazines, 57
    Goat owners, 57
    Livestock auctions, 58
    Online, 57
French Alpine (see Alpine goat)

G

Goat anatomy, 18
Goat breed, 26
    Milk, 26
    Meat, 26
    Fiber, 26
Goat breeding, 152
    Birthing, 166
    Breeding, 163
    Colostrum, 168
    Doe's cycle, 156
    Keeping doe and kid records, 175
    Nursing, 169
    Prebreeding practices, 162
    Pregnant doe, 164
    Suitable buck, 160
    When to breed, 153
Goat cheese, 7
Goat dairy, 32, 34
Goat farmer, xi
Goat fiber, 11, 32
    Angora, 11, 14, 49, 209
    Cashmere, 11, 14, 49, 208
    Mohair, 32
Goat grooming, 180
    Castration, 184
    Disbudding, 186
    Hoof trim, 180
Goat meat, 11, 32, 34, 44
    Cabrito, 11, 45
    Chevon, 11
Goat ownership, 10
Goat production, 207
    Fiber collection, 208, 209
Goat terminology, 20, 62
Goat, 4, 14
    Mood, 19
    Pet, 7
    Provide lesson for responsibility, 7
    Well-being, 19
    Herd animals, 28
    Companion animals, 28
Goat's milk, 4, 7, 191
    Calcium, 11
    Dairy doe, 194
    High butterfat content, 194
    Lower cholesterol, 11
    Makes best cheese, 194
    Making of milk, 197
    Milking equipment and supplies, 178
    Naturally homogenized, 192
    Niacin, 11
    Phosphorous, 11
    Raw versus pasteurized milk, 195
    Riboflavin, 11
    Soap, 11
        Eczema, 11
        Minor irritations, 11
        Skin rashes, 11
    Treats insomnia, 11
    Treats vomiting, 11
    Vitamin A, 11
    Vitamin B, 11

Goats versus sheep, 22
Grade goat, 27
Guardian dogs, 91
   Akbash, 92
   Anatolian shepherd, 92
   Great Pyrenees, 92
   Karst, 92
   Komondor, 92,
   Kuvasz, 92
   Maremma, 92
   Ovcharka, 92
   Shar Planinetz, 92
   Slovac Cuvac, 92
   Tatra Sheepdog, 92

H
Housing a goat, 71
   Converted shed, 75
   Extreme weather, 73
   Floors and beddings, 78
   Grain storage, 80
   Indoor manger, 77
   Lighting, 79
   Water system, 79

K
Keeping goat healthy, 126
   Annual testing and veterinarian visit, 127
   Common diseases and illnesses, 136
   Drenching, 131
   First aid, 132
   Health records, 133
   Poisonous plant ingestion, 146
   Signs of parasitic infestation, 129
   Vaccination and worming, 134
   Weighing goats, 129

L
LaMancha goat, 38, 40
Livestock Guardian Dogs, 92

M
Milking a goat, 200
Miniature breed, 51
   Kinder, 27, 51
   Mini Mancha, 51
   Mini Nubian, 51
   Pygora, 27, 51
Myotonic goat, 45, 48

N
Natural herd protectors, 93
   Donkeys, 93
   Guard dogs, 93
   Llamas, 93
Nervous (see Myotonic goat)
New Zealand Kiko goat, 45, 46
Nigerian Dwarf goat, 32, 51
Nubian goat, 27, 41

O
Oberhasli Brienzer (see Oberhasli goat)
Oberhasli goat, 30, 42

P
Paperwork, 62
Purebred Boer Buck, 46
Pygmy (see African Pygmy)
Pygmy milk, 53
   Calcium, 53
   Iron, 53
   Lower in sodium, 53
   Phosphorous, 53
   Potassium, 53

R
Raising goat, 7
   Cut purchase for milk and cheese, 7
   Fridge full of meat, 7
No need to buy fertilizer, 7
Recipes, 215
   Calzone with Goat Cheese, 218
   Curried Goat, 220
   Fast Soft Goat Cheese, 216
   Goat Milk Feta Cheese, 217
   Homemade cheese, 215
   Honey and Oatmeal soap, 220
Rock Alpine (see Alpine goat)

S
Saanen and Sable goat, 38, 42
Savannah breed, 45, 48
Scrub goat, 27
Selection criteria, 26
   Inquisitive and attentive eyes, 26
   Medium to high level of physical energy, 26
   Willingness to please, 26
South African Boer goat, 45
Spanish goat, 45, 47
Star milker system, 67
Stiff Leg (see Myotonic goat)
Swiss Alpine (see Oberhasli goat)

T
Tennessee Scare Goat (see Myotonic goat)
Toggenburg goat, 38, 43
Transporting the goat, 64

V
Vitamins and Minerals, 119

W
When buying goats, 60
   Age, 60
   Ancestry, 60
   Birth date, 60
   Health, 60
   Production, 60
   Temperament, 60
   Veterinarian visits, 60
Wide area for goat, 84
   Away from predators, 89
   Fencing, 86
   Gate, 87
   Pasture, 88
Wooden Leg (see Myotonic goat)